THE ALZHEIMER'S CATASTROPHE by Cynthia Janus, MD is an important first choice resource for anyone who finds themselves swimming in the worry and wonder of what to do for a loved one or friend who has been diagnosed with dementia or Alzheimer's. Penned from first-hand experience attempting to provide the best care for her husband Sam, Cynthia Janus becomes more and more frustrated and angry with the lack of scientific knowledge and effective treatment available to deal with this debilitating disease. Her frustration is the impetus for her exhaustive search for answers which unfortunately remain woefully lacking. If someone is looking to come to knowledge about the current state of this debilitating disease, I highly recommend this book which cuts through all of the noise and false claims in an understandable manner.

William F. McCarthy, PhD, RN
Professor Emeritus – State University of New York -- Corning

Dr. Janus' account of her experiences as a caretaker for an Alzheimer's patient, her late husband, is a heart rending story. As a physician, she has been on both sides of the giving and receiving of assistance from the medical establishment. Her book is a plea to improve the services available to the Alzheimer's patient as well as that patient's caretaker. This book clearly illustrates that only with a combination of focused research, dedicated health workers and support from the medical establishment for patients and caretakers alike, can we hope to make any progress in fighting and managing the scourge that is Alzheimer's.

Stan Keyles, Psy.D.
Psychologist

Research and medical care for other disease processes like cancer and diabetes are well defined – but Alzheimer's is a maze, with no clear path. Dr. Janus' book helps the caregiver navigate "the maze" with understanding and clarity, and hopefully aids in better decision making. Help in decision making is what it is all about. If you, as the caregiver, do not make decisions, believe me, the juggernaut (insurance, rules, Medicare, SNFs, Home Health, etc.) will make the decisions for you. It is better to be informed, instead of being guided by the mountains of ignorance out there.

Janet Rice, MS, CCC-SLP
Speech-Language Pathologist

For those of us whose relatives, friends and colleagues have faced the devastating toll that Alzheimer's disease takes, this book offers an invaluable and comprehensive overview. Dr. Janus has provided us with a well written and realistic guide to understanding Alzheimer's. This easy-to-read book addresses not only issues related to patients, but shares personal family stories and the current status of treatment and long term care. This book is an excellent resource for those dealing with this tragic illness.

J. Leibman, M.D.

THE ALZHEIMER'S CATASTROPHE

THE LONG UPHILL BATTLE AGAINST ALZHEIMER'S DISEASE AND WHY WE CAN'T AFFORD TO LOSE

CYNTHIA JANUS, M.D.

Copyright © 2016 Cynthia Janus, M.D.

All rights reserved. No part of this book may be reproduced, stored, or transmitted by any means—whether auditory, graphic, mechanical, or electronic—without written permission of both publisher and author, except in the case of brief excerpts used in critical articles and reviews. Unauthorized reproduction of any part of this work is illegal and is punishable by law.

ISBN: 978-1-4834-4357-7 (sc)
ISBN: 978-1-4834-4356-0 (e)

Because of the dynamic nature of the Internet, any web addresses or links contained in this book may have changed since publication and may no longer be valid. The views expressed in this work are solely those of the author and do not necessarily reflect the views of the publisher, and the publisher hereby disclaims any responsibility for them.

Any people depicted in stock imagery provided by Thinkstock are models, and such images are being used for illustrative purposes only.
Certain stock imagery © Thinkstock.

Lulu Publishing Services rev. date: 02/02/2016

To Sam

CONTENTS

Preface ix
Acknowledgments xi
Introduction xiii

I.   BASIC FACTS 1
II.   MEDICAL AND RESEARCH PERSPECTIVES 11
III. CONTROVERSY AND UNCERTAINTY............. 44
IV. DO WE KNOW WHY SOME PEOPLE ARE MORE VULNERABLE? 55
V.   LONG TERM CARE AND NURSING HOMES.... 77
VI. FINAL THOUGHTS.. 107

About The Author 117
Endnotes 119

PREFACE

The subject matter of **The Alzheimer's Catastrophe** is the status of the fight to end the relentless devastation caused by Alzheimer's disease on affected individuals and their loved ones. The lack of a true scientific breakthrough that would lead to an effective treatment and the considerable deficiencies in long term care have resulted in significant stress and hardship, as well as a much diminished quality of life, for people with Alzheimer's disease and their caretakers.

Several years ago I lost my husband to Alzheimer's. Many of the questions I had about the disease could not be answered and I felt compelled to learn more. What was happening in the brain of someone with Alzheimer's disease to make things go so wrong? What causes these changes? What makes some people get Alzheimer's and not others? Could anything have been done to prevent the dire consequences of the disease? Are there any more effective treatments now? Are there any viable new medications in the "pipeline"? I also wondered about what was really going on in those dreaded nursing homes and other similar type long term care facilities I heard other caregivers complain about. Aren't there any better alternatives for taking care of those affected? Why have pharmaceutical companies, care facilities, health agencies and others in a position to help failed to provide any meaningful relief for Alzheimer patients and their loved ones? Why is it taking so long?

While my questions are still not fully answered, the **Alzheimer's Catastrophe** is my attempt to share with the reader what I have learned. Better days may yet lie ahead if we understand and spotlight the impediments to progress and we advocate for change.

ACKNOWLEDGMENTS

I wish to express my gratitude to the many loving, patient and exhausted caregivers who shared their individual experiences and personal pain with me. I would like to acknowledge Sharon B. Henrich for her invaluable research assistance. I also want to thank my son, Lazar, for his technical help and his ability to make me laugh even in the face of such weighty and frustrating subject material.

INTRODUCTION

What I Learned About Alzheimer's Disease (The Hard Way)

I am a physician, retired now. Practicing my specialty, radiology, for some 35 years, I felt gratified that imaging findings helped physicians in other specialties diagnose and plan a course of treatment for their patients on a daily basis. I was fascinated by the seemingly endless number of new ways to treat and manage patients with all kinds of illnesses and injuries and I was particularly proud of the new technologies in my field of radiology which made it possible to image patients more safely and more quickly. Radiologists with special expertise increasingly used new techniques to treat certain disorders, eliminating the need for surgery in many cases. Of course, medicine has always had some treatment failures, but with our vast array of scientific tools and medical knowledge, there was always hope that we would prevail in the end.

Then, I joined the growing ranks of those who have lost a loved one to Alzheimer's disease. The actual physical loss comes after a long period of helplessly watching the person you care about gradually lose their memories, talents, and all other mental capacities until eventually their core personality seems to disappear. The mental confusion and disabilities and, later on, the accompanying physical decline go on progressively and relentlessly over years for those affected by the disease and also take an immense toll on their caretakers. I learned how very little medicine has to offer in alleviating their distress. Adding "insult to injury" was the discovery that there is money to be made off the suffering by some pharmaceutical companies, lawyers, care facilities and caregiver organizations. Don't get me wrong. For the most part, these institutions and services can be vital and there are many well meaning organizations striving to help those with AD.* However, some of these entities operate only out of greed and the fallout on patients and their families can be devastating.

We, the general public, have not been getting the whole story. We hear hopeful news of the discovery of new substances or of other new "breakthroughs" that *may* lead to a better understanding of the disease and new drug trials that *might* lead to a more effective treatment of patients with Alzheimer's. We read of ambitious projects involving brain mapping and gene mapping. Health care organizations show us images in magazines and on T.V. of smiling, elderly patients with their loving, paid caretakers. Family members of Alzheimer patients are encouraged to join support groups, take care of themselves and find outlets to lessen their stress. Believe me, these goals are a lot harder to achieve than they sound. For me and the many other families I have spoken to, the optimistic news items and lovely images are far from reality.

True, there have been impressive improvements in imaging the brain and documenting anatomical changes which have been associated with Alzheimer's. Medical personnel have gained more information about, and experience with, the disease with which to advise families on what to expect in terms of symptom progression and they can offer help with some coping strategies. However, at this time, there is NO CURE or even effective long lasting treatment for Alzheimer's disease. In fact, no one really knows for sure the exact processes that occur in the brain, anatomically and microscopically, that lead to the development of the clinical symptoms of

---

* Sometimes, the term Alzheimer's disease will be replaced by the name Alzheimer, or Alzheimer's or by the initials AD.

Alzheimer's disease. In the end, it's really just the patient struggling against this formidable disease. In the end, THE DISEASE ALWAYS WINS.

I hope that someone will find a way to cure, or even better, prevent Alzheimer's disease but that may take a long time. Even if a cure is found tomorrow, it will take years or decades to determine if there are real long term benefits and to ensure that there are no severe harmful effects. In the meantime, what we can and should do is to provide a better quality of life for Alzheimer patients. This could involve strategies to allow people to remain in their homes as long as possible by providing financial aid and other help to family caretakers. There should be special facilities manned by medical personnel who are knowledgeable and well trained in the area of dementia for those people whose symptoms are so severe that living at home is dangerous for themselves or their family, or who have serious associated medical conditions.

Currently, those with more advanced stage Alzheimer's disease often end up in nursing homes or similar type long term care facilities. These places are frequently understaffed. Nursing assistants and other employees may have little to no formal training or prior experience to help them in the challenging task of caring for these complex patients. People with Alzheimer's disease, especially the later stages, generally have no voice in their care. Typically, they are not treated as individuals with specific needs but are forced into a regimented day that fits the needs of the facility in which they live. Any attempt at resistance on their part is quashed, usually with medication. They do not deserve the indignities and, sometimes abuse, offered up.

As a result of my long journey into the frightening world of Alzheimer's disease, I have continued to follow the touted breakthroughs and, more often, failures of multiple proposed treatments and the numerous theories about what makes people vulnerable to the disease. I have tried to fathom the complex problems and deficiencies of nursing homes and other such facilities that care for these patients.

There have been books written about the scientific aspects of Alzheimer's disease, numerous good caregiver guides and many firsthand accounts of living with the disease written by truly brave and inspiring patients or their loved ones. **The Alzheimer's Catastrophe** is not meant to be a medical treatise or a self help book but rather a realistic look at the devastating toll that Alzheimer's disease takes on the individual, his or her personal world, and ultimately on the rest of us. It is meant to provide an overview of what scientists have learned so far, what directions research has been taking and what deficiencies have been revealed in the overall approach to long term care of Alzheimer patients. This information is compiled from scientific articles in medical journals or on the internet, news media. personal experience, and from other caregivers, the true "experts in the field." *

In the end, I have learned that providing effective medical treatment and a better quality of life for people with Alzheimer's disease may be a herculean task but for those touched by the disease, help cannot come soon enough.

---

* The names and identities of the people quoted or described in the anecdotes have been changed.

I BASIC FACTS
Defining the Problem

A prominent feature of Alzheimer's disease is *dementia* and sometimes these terms are used interchangeably. However, dementia is not really a specific disease but rather a general term for a group of symptoms. Simply stated, brain cells are damaged and lose the ability to communicate with each other, causing the brain to malfunction. For example, a person with dementia will suffer loss of memory. Their reasoning and judgment may be impaired and they may display inappropriate behavior. They can lose the ability to communicate and be disoriented as to time, place and person. Difficulty in planning and inability to think abstractly are other symptoms. People with dementia may neglect their personal appearance and may be oblivious to safety issues. They may become unusually confused, suspicious or fearful in a new environment. Later symptoms include agitation, paranoia and even hallucinations. If you interact with a person with dementia you may find that they have difficulty in performing routine tasks or that they no longer recognize people they used to know. They may get lost easily and not be able to follow previously well known routes or to follow directions. They may repeat the same questions over and over again.[1]

As the dementia progresses, new symptoms bring great challenges to the caretakers of loved ones with dementia who don't know what to expect from one day to another. Stories like the following abound:

Cynthia Janus, M.D.

Tom: *My wife, Marie, was diagnosed with Alzheimer's. Her sister looked in on her every day, so I felt fairly comfortable when I was at work. Marie was not supposed to take her car out but wait for her sister to come if she wanted to go out. One day, while I was at work, I got a call from the police, telling me that they found my wife wandering around in a small strip mall. I nearly had a heart attack. She didn't know where she was. Luckily, one of the store owners saw that something wasn't right and called the cops. They found my name and phone number in her purse. I knew then and there that things were going downhill and took away her car keys.*

Jean: *My husband Bob's dementia was getting really bad. He didn't seem to understand what we were saying and it was getting harder for us to understand him. He got things mixed up and used words that weren't real words. He didn't recognize his grandchildren and after awhile they stopped asking for him on the phone which made me feel terrible.*

Phil: *I was up all night with my wife Alice, who kept insisting that it was time to go home. She didn't know she WAS home. I was really tired. The next thing I remember was that the dog came over for me to walk him. I looked at the clock. It was seven A.M. Then suddenly I realized Alice wasn't in bed. I ran around the house frantically looking for her and praying she didn't get out. Finally, I found her sitting on a box in the garage.*

Signs of dementia have sometimes been considered an inevitable consequence of old age or "senility." NOT TRUE. There are plenty of elderly people who manage to stay "sharp." While many other individuals do have gradual and mild memory loss as they age, the symptoms are not so severe as to render them unable to perform normal daily functions as seen in people with true dementia. DEMENTIA IS NOT A NORMAL CONSEQUENCE OF AGING.

There are many conditions that can bring about the symptoms of dementia.[1] Alzheimer's disease is the most common form of dementia.[2] Reported statistics vary but, according to most reports, Alzheimer's accounts for up to 60-70% of cases.[1,2] Most cases of Alzheimer's disease occur in people 65 years of age and up. In a very small percentage of cases (probably up to 5%), AD can occur in people in their 40's and 50's and this is known as early onset or younger onset AD. The second most common form of dementia is vascular dementia which can be brought about by a series of small strokes. Also, these two types of dementia can occur in the same person as they age.

Other diseases associated with dementia include Parkinson's disease, Pick's disease, Creutzfeldt-Jacob disease, AIDS, Huntington's disease and Lewy Body dementia. Sometimes, symptoms of dementia can be seen in patients with kidney and liver disorders and brain tumors and even in people with depression. Brain function and memory can also be adversely affected by thyroid problems, vitamin deficiencies, side effects of medications and excessive alcohol ingestion. In these cases the dementia may be treatable so it's important to know what you're dealing with. This book deals predominantly with Alzheimer's disease, although some of the material is applicable to people with other types of dementia.

How Does the Doctor Diagnose Dementia?

Doctors start off as they would in evaluating for any disorder by taking a careful history and doing a physical exam. There may be characteristic behavioral and laboratory changes to help identify the specific kind of dementia. On the other hand, findings in the different types of dementia may also overlap, making it difficult to identify the specific type. However, a particular type of dementia may be associated with changes in the cells in specific areas of the brain which can be seen on imaging exams.

What is Different About Alzheimer's Disease?

Two types of abnormal proteins have been associated with development of Alzheimer's disease.[3,4,5] **Beta amyloid** plaques are deposits of a protein fragment that accumulate in spaces between neurons, which are the brain's nerve cells. (The brain has on the order of 100 billion nerve cells. It's pretty amazing how well brain processes are integrated and coordinated the vast majority of the time.) Twisted fibers of another protein called **tau** build up inside the nerve cells. These plaques and tangles of abnormal proteins are found to be much greater in amount in people with Alzheimer's disease compared to people of similar age without the disease. They also spread in a characteristic pattern through the brain that has come to be associated with AD. For instance, early on the abnormalities are found near and in an area of the brain called the hippocampus, a center of learning and memory, and then spread to other regions. Scientists have thought that the plaques and tangles block communication among nerve cells in the brain. This disrupts vital processes that the cells depend upon to survive. The disruption and death of nerve cells in various parts of the brain causes the memory loss and other problems controlled by these areas. The method in which Alzheimer's disease develops is not entirely known and the role of amyloid and tau proteins has been questioned by some.

Mild Cognitive Impairment: What is It?

A person with mild cognitive impairment may have problems with memory or other functions of the brain. However, these people can still function independently in their daily lives.[6] Many people with mild cognitive impairment will go on to develop Alzheimer's disease or some other type of dementia while symptoms may remain stable or even improve in others. The doctor may determine that a patient has mild cognitive impairment on the basis of medical history and information about how he or she is functioning. A physical, including neurologic exam, assessment of mental state and mood are part of the evaluation, as well as laboratory tests and imaging exams, as needed. Much of the evaluation for mild cognitive impairment and dementia can be performed by the general physician but patients are usually referred to neurologists and other medical specialists for specific testing.

What are the Stats?

Attention has recently been given to the fact that Alzheimer's disease and other dementias affect a very large and growing number of people in the United States and throughout the world. Statistics vary slightly from source to source but they all tell the same story.

WORLDWIDE: In 2012, the World Health Organization (WHO) reported that 35.6 million people throughout the world have dementia of some type, including Alzheimer's disease.[2] Slightly more than half of these people live in low and middle income countries. The number of people affected was expected to be on the order of 65.7 million in 2030 and 115.4 million in 2050. Much of the projected increase in numbers is related to the increasing number of people with dementia reported in the low and middle income countries. It appears to be a case of both better reporting in some areas of the world and a true increase in the number of people with dementia. Statistics cited by the organization Alzheimer's Disease International include an estimated 44 million people with dementia worldwide in 2013 compared to a 2010 estimate of 35 million. According to this organization, the number of people with dementia is expected to reach 76 million in 2030. By 2050, that number is expected to climb to 135 million.[7]

Reading the statistics can sometimes be confusing. Results can vary according to the criteria and database used in the study. Although there are some differences in the estimates given by various organizations, all sources report an expected marked rise in the number of people with dementia in the years to come.

UNITED STATES: According to the Alzheimer's Association's Latest Facts and Figures Report for 2015, 5.1 million people, 65 year and older, in the United States have Alzheimer's disease, including 3.2 million women and 1.9 million men. An additional 200,000 individuals are younger than 65 years of age and have the early onset form.[8]

In 2013, the New York Times reported that according to a study by the RAND (Research and Development) Corporation, 15% of people who are 71 years of age or older have dementia. This represents approximately 3.8 million people. By 2040, there would be an increase in numbers affected to 9.1 million.[9] The research was based on information collected for almost 10 years on 11,000 people taken from a large database called the Health and Retirement Study. Besides those people with actual dementia, it had previously been estimated from the same data that 22% of people 71 years of age and above have mild cognitive impairment. It was expected that 12% of these individuals would progress to the dementia category each year.[9] A report by the Association of Health Care Journalists[10] describes the findings of a RAND study in 2014 which focused again on the growing number of people with Alzheimer's disease and also the need for improvement in long term support services. It cited the expectation for the number of Americans with AD to be about 14 million by 2050, creating a huge strain on the long term services associated with dementia care and support.[10]

The Alzheimer's Organization gives similar predictions in their 2015 report. With so many more individuals living beyond age 65, the number of people diagnosed with Alzheimer's is expected to reach approximately 13.8 million in 2050.[8] This figure represents almost triple the number of people with AD in 2015. Again, there is no good news in any of these statistics.

STATISTICS ON MORTALITY: The death toll from Alzheimer's disease is also disheartening with about 700,000 deaths from AD expected in 2015.[8] Along with the ever growing number of people diagnosed with Alzheimer's, the number of those dying from the disease has, of course, also increased. According to statistics from the Alzheimer's Association, there was a 71% increase in the number of deaths in 2013 when compared to the mortality rate in the year 2000.[8] While the mortality rate from many diseases has fallen, the number of people dying from AD keeps rising.

You may be wondering how the death rate from Alzheimer's disease compares with other diseases. According to the Centers for Disease Control and Prevention, Alzheimer's disease is the **sixth leading cause of death**.[11] However, according to a recent report of a study by researchers at Rush Alzheimer's Disease Center in Chicago, AD could really be ranked as the **third** leading cause of death after heart disease and cancer.[12] This is because the cause of death listed on death certificates does not often accurately reflect the contribution of AD. For example, while approximately 84,000 deaths were attributed to AD in 2010 by the CDC, the number is really close to 500,000 among those 75 years of age and up, according to the report. It may take up to 10 years to die from the effects of Alzheimer's disease. The person with AD, for example, may be more likely to die from pneumonia and other infections. Many have trouble swallowing as part of the disease which can lead to aspiration of food. When this happens, food enters the trachea (windpipe) instead of the esophagus (foodpipe) which can result in serious pneumonia. The death certificate, however, may only cite the pneumonia, the last fatal complication, instead of Alzheimer's disease itself as the cause of death.[13,14]

STATISTICS FOR WOMEN: Of the five million Americans with Alzheimer's disease, two-thirds are women.[8] A woman 60 years of age or over has a 1 in 6 chance of getting Alzheimer's disease in her lifetime (much greater than the risk of getting breast cancer) while the chance of developing it for a man is 1 in 11.[15] The reason for this is not known but some researchers are studying the possible effects of hormonal differences. Women also make up the majority (approximately two-thirds) of the caregivers of those with Alzheimer's disease.[8] People with AD will ultimately need round the clock care for a variable period of time which can often go on for many years. Because of the heavy burden of care-giving, many women change from full-time to part-time employment status and some just leave the work place.

FINANCIAL STATS: Research by an economist at the Rand Corporation, and financed by the federal government, lead to the finding that the cost for dementia care for Americans is at least as high as that for heart disease and cancer.[16] According to the study, direct health care expenses for dementia in 2010, including nursing home placement, was $109 billion dollars. By comparison, the cost was $102 billion dollars for heart disease and $77 billion dollars for cancer care.[16]

There was also a considerable amount of what is called "informal" care for dementia given at home. Thus, while the cost of dementia care in 2010 was approximately $109 billion, the price tag was $159 to $215 billion when you include the monetary value of informal care.[16] There have been no detailed studies of the amount of money spent on informal care at home for people with heart disease and cancer. However, it is assumed that, because of the nature of Alzheimer's disease and the greater need for care and close supervision of these patients, the monetary value for informal care in the case of AD probably exceeds that of heart disease and cancer. The researchers pointed out that people with dementia are not receiving significantly more health care services, but rather it is the nursing home costs and the formal and informal home care costs that make up 75% to 84% of the price tag.[16]

The long term costs, in general, are very high.[17,18,19,20] The RAND researchers have estimated that the total costs of dementia care, which ranged from $159 billion to $215 billion in 2010, could more than double by the year 2040.[10] The Alzheimer's Association gives similar estimates for the current cost of dementia care. They estimated a direct cost of $226 billion for Alzheimer's care in 2015 with Medicare paying for one half the costs. Their prediction of future costs is higher, reaching $1.1 trillion dollars by 2050.[8]

No matter what statistics you read, the conclusion drawn by all of the studies is that both the number of people with dementia and the cost of care are expected to increase dramatically. All of the calculations are cause for serious concern that our current health care system will be overwhelmed.

That Which Cannot Be Measured

Those who have witnessed a loved one, friend, neighbor or colleague succumb to the ravages of dementia often experience an array of problems and emotions. At first, your relative, friend or colleague seems a little "off their game" and not as sharp as they used to be but you assume that they are just under stress or distracted by some issue going on in their life. They may become forgetful, even when doing something or going somewhere that was once routine and may say things that just don't make sense. Sometimes they withdraw from activities they used to enjoy. You are uneasy and a little worried. You express your concerns to them and they may respond with denial or even anger. They may "laugh it off" or they may express concern themselves.

After some time, as symptoms worsen you come to realize that – yes - they have dementia. Receiving the diagnosis of Alzheimer's disease is a terrible blow because no matter how encouraging and supportive physicians may be, they cannot offer hope of a cure. You start to wonder, "How is this going to go? How will I help my spouse or parent and still go to work?" You ask yourself, "How will I take care of my family and still be there for Mom or Dad? There are all kinds of organizations that will answer my questions and advise me on where to get help, aren't there?"

But wait… life for the elderly with dementia doesn't seem that bad, if media advertising is to believed. We see an elderly man or woman taking a walk or having some other pleasant interaction with their paid caregiver who is calm, smiling and totally engaged with them. I am here to tell you it doesn't always go that way. If you're lucky to have family willing and able to participate, they can take turns with you as your loved one begins to need more help with daily activities. Families may also engage a company to provide home aides to watch over their loved ones but the companies vary in quality and the service is expensive. Many of the aides have little or no training in caring for the elderly, much less those with dementia. In addition, the company may not always provide the same aide for your loved one, making continuity of care or bonding with the aide more difficult.

Meanwhile, while you are trying to cope, your loved one with Alzheimer's disease is experiencing a continual and relentless loss of memory which eventually progresses to not being able to recognize family and close friends and even you. They forget what everyday objects are used for. They appear to undergo a loss of their sense of self. Everyday simple acts that we take for granted like eating, grooming and dressing become insurmountable. There are sleeping irregularities such as sleeping during parts of the day and wakefulness at night. Eating and drinking often become complicated by aspiration of the food and liquids. Invariably, bladder or bowel incontinence appears as the disease progresses. Sometimes, you see frustration, anger or sadness in the eyes of your loved one who cannot tell you what they are feeling. You want to help them but you can't. Confusion, disorientation, wandering, agitation and physical outbursts add to the general mayhem and misery for those affected and their families. The disruption of the normal life of the individual's family and the tremendous toll on their caretakers is well documented.

When families can no longer keep their loved one at home, they may turn to a long term care facility but these vary greatly in quality and service and may not provide the medical care or safety that the AD patient requires. Finally, there is the option of the nursing home or skilled nursing facility. The general perception of the public is that these range from places where you lie around waiting to die to actual horror houses. In the worst instances, these are places where any shred of pleasure or even dignity is finally removed. Indeed, many relatives of nursing home patients and former employees report cases of indifference, neglect and incompetence. It is no wonder that when you ask the average person what they think about nursing homes, most responses will be something on the order of *"I would never put my parents into one of those places,"* or *"I'd rather die than go into a nursing home."*

As you have seen, and may already know from personal experience, things can get pretty bad for the person with Alzheimer's disease or other type dementia. Recently, we have been hearing and reading more about Alzheimer's in the media. How long has this disease been around? How have doctors treated the patient with AD and what attempts have been made to find a cure? Let's take a look back.

II MEDICAL AND RESEARCH PERSPECTIVES

Looking Everywhere for a Cure

Who Is Alzheimer?

Did you ever wonder how Alzheimer's disease got its name? In 1901, Aloysius (Alois) Alzheimer, a German psychiatrist and neuropathologist, observed a 51 year old female patient named Auguste Deter at the Frankfurt Asylum in Germany. This woman had become increasingly unable to care for herself at home. She had trouble reading and writing, and suffered from memory impairment and disorientation. Her symptoms got worse and included hallucinations and gradual loss of higher mental functions. Dr. Alzheimer became very interested in this patient's disease and, when she died in 1906, he had her brain and medical records brought to Munich University where he was working in the laboratory of Emil Kraepelin, a German psychiatrist. Upon anatomic and pathologic exam, the cerebral cortex* of the woman's brain was found to be thinner than normal. Alzheimer described two additional abnormalities consisting of

---

* Cortex – the outermost portion of neural tissue of the brain; also referred to as the gray matter

plaques and neurofibrillary tangles in her cerebral cortex. This formed the basis of his hypothesis that there was an organic cause for her condition.[1] Plaques had previously been found in the brains of elderly people but nerve tangling had not been described. Plaques and tangles are still considered by most scientists to be key elements in the pathology of Alzheimer's disease.[2]

Dr. Kraepelin, known for his classification of schizophrenia, incorporated the study of Frau Auguste D in the eighth edition of his textbook "Psychiatrie" in 1910. In his theories on mental illness, he talked about the role of organic changes occurring in the brain. He described "Alzheimer's disease" as a set of distinct symptoms and clinical manifestations caused by specific organic abnormalities.

Pretty good work, I'd say, since Dr. Alzheimer didn't have all the advanced scientific equipment that we have today at his disposal or even a Federal grant! On the other hand, he probably didn't have all the bureaucratic burdens and pressures that scientists currently have to put up with. In any case, what progress have we made since Dr. Alzheimer put forth his findings?

What Medical Progress Have We Made in Finding Treatments?

There have been many attempts over the years to develop medications to stop, or at least slow, the changes in the brain that result in Alzheimer's disease. In order to follow the progress of the researchers and pharmaceutical companies, first consider the following. Over time, people with Alzheimer's disease make less of a brain chemical called acetylcholine which helps thought processes and memory. Many of the medicines prescribed for AD patients are in the class of drugs called **cholinesterase inhibitors.** The drugs improve the functioning of brain cells by inhibiting the enzyme acetylcholinesterase which breaks down acetylcholine.[3,4] However, these medications cannot make acetylcholine and they cannot totally stop or reverse the loss of acetylcholine and the destruction of brain cells. Generally, it has been found that these drugs may delay the worsening of Alzheimer symptoms for, on average, six to twelve months for about one-half of the people taking them, but then, after a time, stop working.[4,5,6]

In 1993, the first of these cholinesterase inhibitors, a medication called Cognex(Tacrine), was developed in an attempt to slow the progress of Alzheimer's. It is no longer used because of its serious side effects including liver damage. More recently, other drugs have been produced and approved by the FDA (Table 1), but the consensus opinion by experts still holds that they may only temporarily slow the progression of symptoms and they are not effective in all patients.[4] Aricept (donepezil) is the only cholinesterase inhibitor approved by the FDA to treat all stages of AD. Exelon (rivastigmine) and Razadyne (galantamine) were approved to treat mild to moderate AD. Namenda (memantine) was approved by the FDA for treatment of moderate to severe AD. It is involved in the regulation of glutamate which is another chemical involved in memory and learning. Namenda can be used alone or together with other AD medications. Like the other medications, it may delay worsening of symptoms for some AD patients. Both classes of drugs also have side effects including nausea, vomiting, loss of appetite and increase in the number of bowel movements in the case of the cholinesterase inhibitor drugs, and headache, constipation, confusion and dizziness in the case of Namenda.[3,5]

More recently, in 2014, the FDA also approved a drug called Namzaric to treat patients with moderate to severe dementia. (Table1). It combines memantine, the key ingredient in Namenda, and donepezil, which is found in Aricept. Namzaric was introduced in the United States in a two dosage strength in 2015.

TABLE 1

FDA Approved Medications for Alzheimer's Disease

Method of Action: Cholinesterase Inhibitors

Brand Name	Drug Name	Date FDA approved	AD stage
Cognex	tacrine	1993	mild-moderate
Aricept	donepezil	1996	all stages
Exelon	rivastigmine	2000	mild-moderate
Razadyne	galantamine	2001	mild-moderate

Method of Action: Involved in the Regulation of Glutamate

Brand Name	Drug Name	Date FDA approved	AD stage
Namenda	memantine	2003	moderate-severe

New Combination

Namzaric	donepezil & memantine	2014	moderate-severe

Vitamin E has also been tried in the past to manage the symptoms of Alzheimer's disease. It is in the class of anti-oxidants with the potential of protecting the body's tissues, including brain cells, from the harmful effects of certain chemicals. A study in 1997[7] indicated some temporary beneficial effects. Since then, other investigations[8] showed that, at high doses, vitamin E can have a negative interaction with other drugs and may slightly increase the risk of death, especially in individuals who already have coronary artery disease.

More than 100 years following the work of Dr. Alzheimer, we still have no treatment to appreciably reverse the debilitating symptoms or to slow down the progression of the disease named after him. There is no way to accurately test individuals for susceptibility to Alzheimer's dementia. The precise chain of events that causes the brain changes described by Alzheimer, and the reason or reasons for their development remain unknown.

Jack: *Every time I took Dad to his neurologist appointment, there would always be an attractive sales rep with her luggage cart packed with meds and her arms filled with promotional brochures. Those drug companies must have been making out like bandits but I didn't see any improvement when Dad was on the drug.*

Sometimes It's All About the Money

You may be surprised, as I was, to learn about some of the maneuvering and controversy that may accompany the development and marketing of a new medication, including those intended for individuals with Alzheimer's disease. Take, for example, the drug Aricept (donepezil) which had been approved in 1996 for use in patients with Alzheimer's disease and which was widely advertised. It was eventually shown that any patient improvement was likely to be temporary and many said that any changes for the better were really gone by six to twelve months. Nevertheless, caretakers were desperate for any medication with the potential to slow the inexorable decline in their loved ones. Print and TV ads fueled their hopes. Doctors, too, who had little else to offer, prescribed Aricept (and later, other medications like it) for their patients. The result was that, since its approval in 1996, sales of Aricept produced over two billion dollars annually. However, it was about to lose its patent protection in November 2010, which would result in competition from cheaper generic versions of the drug.

Not to worry. In 2009, four months before the patent production would have run out, the Japanese pharmaceutical company Eisai applied for FDA approval of a 23mg version of Aricept, a dosage that can't be attained by combining the 5mg and 10 mg doses of the drug, available in the generic form. The medication, Aricept 23, was developed by Eisai, which along with the drug company Pfizer, marketed it in the United States.[9] The manufacturers received approval for the higher dose of the medication that extended their exclusive right to sell the drug.

The medication was supposed to improve both cognitive ability and overall functioning in patients. However, it seems that the higher dosage worked only slightly better but had potentially dangerous side effects. A clinical trial which followed 1,400 patients showed that the 23mg, higher dosage, of Aricept yielded slightly better results on tests of cognition, but failed to meet the additional goal of improved global function in people with moderate to severe AD. It had no impact on their degree of functioning in everyday life, as reported by their caregivers. In addition, the higher dosage resulted in a significant increase in gastrointestinal symptoms like nausea and vomiting which could be dangerous for the elderly.[9] Although, the recommendations of both a clinical and a statistical reviewer for the FDA were not to approve the higher dosage of Aricept, it ultimately got approved in July 2010.[9]

The story does not end there, however. In May 2011, an organization called the Public Citizen's Health Research Group asked the FDA to remove Aricept 23 from the market stating that, while subjecting patients to more side effects, the higher dose drug was not shown to afford any significant improvement to their general condition.[10,11] The Public Citizen Group did not receive any answer to their petition for removal of the drug and they filed a lawsuit against the FDA in September 2011. In November 2011, their petition was rejected. In the complaint, manufacturers were accused of falsely claiming that the drug had improved both cognitive function and overall or global functioning in advertisements to physicians and on the drug label. The drug label information was corrected and, despite the above concerns, the drug was kept on the market.

Apparently, pharmaceutical companies often use interesting approaches to extend their exclusive rights to sell a drug and remove competition from generic drug makers. They may alter their medication by changing the chemistry slightly or make an extended release version. The search for a treatment to improve the lives of millions of people suffering with Alzheimer's disease is the goal of most researchers. However, to state the obvious, the lure of financial rewards can readily overshadow more noble objectives.

Even the newer drugs hold no promise of a cure. As you recall from earlier in this chapter, Namenda is another one of the medications used in patients with Alzheimer's disease. I remember seeing an ad in medical journals for Namenda XR (extended release) stating that "In moderate to severe Alzheimer's disease once-daily Namenda XR 28mgs + AchEI (stands for acetylcholinesterase inhibitor, discussed earlier) demonstrated improvements in cognition and global function." The ad reminded us that the drug would "Help slow symptom progression…Because there's so much to lose." A picture showed the back of a well-dressed gentleman with thinning gray hair, looking at the smiling faces in family photos of various significant life events: a marriage, a graduation and so forth. At least from the back, he looked really neatly attired for someone with moderate to severe AD. I thought, *"Is he really the patient? It looks like he should be the doctor. Is the medication that good?"* The ad went on to instruct how patients can be switched to the extended release capsules. It advised caution when prescribing this drug to people with liver or kidney impairment and noted other potential side effects, especially headache, diarrhea and dizziness. It reported the results of a "24 week study of 677 outpatients with moderate to severe AD on stable AchEI therapy" in which "adding Namenda XR 28 mgs was statistically superior to placebo* + AchEI" in achieving better cognition and global function, as determined by criteria specified in the ad.

This sounds hopeful, but what happens after 24 weeks? Most importantly was the statement that "There is no evidence that Namenda XR or an AchEI prevents or slows down the underlying disease process in patients with Alzheimer's disease." Sadly, that last sentence is the "bottom line" and one of the few things known with certainty.

It should be noted that the Attorney General of New York State filed a lawsuit against the drug manufacturing company Actavis PLS because it was planning to stop selling the form of the drug Namenda that is taken twice a day. The suit stated that the company was forcing Alzheimer's patients to switch to the once daily Namenda XR.[12] The maneuver was believed to have allowed Actavis to build up a base of prescriptions for Namenda XR and reduce their competition because low cost generics were soon to become available. Actavis countered that the new pill is more convenient and the capsule contents can be sprinkled on food, a benefit for patients who have trouble swallowing. However, according to a decision

---

* Placebo - an inert or innocuous substance, used in controlled experiments studying the efficacy of another substance

by the Second U.S. Circuit Court of Appeals in New York, Actavis was prevented from removing the older Namenda until after the generic forms of the original pill became available. The decision was considered a win for Alzheimer disease patients who would still be able to purchase the original version of Namenda at lower cost.

When Risks Outweigh Benefits

Although it has been established that drugs used to treat Alzheimer's symptoms have limited benefit, a considerable number of patients in the advanced stages of the disease are still prescribed these medications, as well as other drugs, such as cholesterol lowering statins. In one study, the medical records of 5,406 nursing home residents with advanced dementia in 460 facilities, during the period from Oct. 1, 2009 to Sept. 30, 2010, were reviewed. The majority of the patients were severely affected, having lost the ability to speak, recognize family members and walk without assistance. The investigators found that 54% of these individuals were continuing to receive at least one of their earlier prescribed medications that were no longer of any benefit to them. Thus, the patients were still subject to any unpleasant or dangerous side effects like nausea, urinary retention (not being able to empty your bladder), irregular heartbeat and fainting, but they were unable to inform their doctors or nurses about any ill effects they might be feeling. Most patients with advanced dementia have trouble swallowing and are vulnerable to aspiration so taking these unnecessary medications certainly does not help. The practice of continuing to prescribe these drugs also drives up health care costs.[13,14]

The data from the study was also broken down in various ways and analyzed further. For example, the questionable prescribing practices were most frequent in the southern and central portions of both the eastern and western United States. It was lowest in the mid-Atlantic states, New England and the mountain states in the west. Latino patients with dementia were more likely to be prescribed the medications and rates for non-Latino blacks and whites were approximately the same. Men were also more likely to get the drugs than women. The "probably not helpful" medications were also given to hospice patients although they were less apt to receive the drugs than patients outside the hospice system.

The message here is that caretakers of patients with Alzheimer's disease and other dementias should review what medications their family member with advanced disease are taking. They need to know if the drugs are still of therapeutic value or just subjecting their loved one to uncomfortable side effects and unnecessary medical risks.

Strategies Initiated to Address a Growing Problem

While patients with Alzheimer's disease and their families have been, for the most part, silently struggling throughout the years, the formal recognition by Federal agencies of the need for some type of help or intervention emerged slowly. In 1974, the National Institute on Aging (NIA), which supports Alzheimer's research, was established. In 1980, the Alzheimer's Association was formed. In 1983, Congress designated November as Alzheimer's Disease Awareness month. Around 1984, a network of Alzheimer's disease centers was set up at leading medical institutions. Other significant events to focus attention on and to counter AD occurred in the 1990's. In 1991, the Alzheimer's Disease Cooperative Study (ADCS) was established by the NIA. It conducted federally funded clinical trials across the nation. It was in 1993 that the FDA approved tacrine (Cognex), the first medicine for AD, although as previously mentioned, it is no longer used. Alzheimer's Disease International proclaimed September 21 as World Alzheimer's Day in 1994. During the next decade, the NIA, along with the Alzheimer's Association, recruited people for the Alzheimer's Disease Genetic Study. In 2008, the International Society to Advance Alzheimer Research and Treatment (ISTAART) was formed. Conferences on AD were held annually. These events at least shed light on Alzheimer's disease and fostered research, although no major breakthroughs resulted to help patients in a significant way.

Recently, there has been increasing recognition of the profound consequences of Alzheimer's disease not only on the affected individual but also on society in general. The first national plan to address Alzheimer's disease was mandated by the National Alzheimer's Project Act (NAPA) which was signed into law in 2011. This plan to address the problem of AD was released in 2012 and set a target date of 2025 to develop effective treatments and methods of prevention.[15] The National Institutes of Health (NIH) announced new studies in early 2013 to be carried out by members of the Alzheimer's Disease Cooperative Study (ADCS) that had been earlier created to investigate and develop treatments for Alzheimer's disease.

Since these announcements, many areas of research have been undertaken. There have been calls for more money to be allocated to research and more pleas for public donations. Advocates of AD research have pointed out that Washington allocates on the order of $6 billion to cancer research and $3 billion to research on HIV/AIDS while at the same time Congress has never approved more than $600 million dollars for annual funding for Alzheimer's research.[15] Congress added $100 million to the National Institute on Aging for AD research and doubled the $100 million going to a program called the BRAIN Initiative (stands for Brain Research through Advancing Innovative Neurotechnologies.) However, some AD advocates say that more than $2 billion annually in research funding is necessary over the next 10 years if the goals of the 2025 deadline are to be met.

Following, you will find short descriptions of many of the major research studies that have been undertaken to better understand the pathology causing Alzheimer's disease with the goal of discovering a treatment, cure or prevention. The discussion is meant to serve as a broad survey of lines of investigation that have been reported about in medical journals and by the media. It is by no means a definitive or totally complete recounting of all the research on Alzheimer's that is in progress. The results of these studies are usually quite interesting and often promising, but sometimes findings have been contradictory and confusing. Most of the results have been preliminary. In any case, I think it is important to see where research money has been going and to look at what new understanding or progress, if any, has been achieved.

More Drug Trials

Scientists and pharmaceutical companies have been experimenting with a variety of newer drugs in an attempt to reverse, or at least halt, the progression of Alzheimer's disease. However, the optimism after the initial reports has so far mostly fizzled out.

At this point in the narrative, it is helpful to take a more detailed look at **amyloid beta (also called beta amyloid)**, **amyloid plaques**, **tau** and **neurofibrillary tangles**. These words have sort of been the language of Alzheimer's disease and the focus of the bulk of the research. Amyloid plaques contain the protein beta amyloid and these plaques are found in the space between the cell bodies of the brain's neurons (nerve cells). The plaques are located in a portion of the brain called the hippocampus, important for processing memories, and in other portions of the cerebral cortex. What about tau? Under normal conditions, structures called microtubules are an important factor for the transport of materials into the cells and the protein tau is also involved in this process. When a person has Alzheimer's disease, the tau is somehow changed in a bad way and starts to clump together, forming the neurofibrillary tangles. Because of these abnormalities, the nerve cells in the brain can't work well or connect properly with other cells and eventually die. Finding beta amyloid in the brains of people with Alzheimer's disease has led to what is called the amyloid cascade hypothesis.[16] This theory postulates that accumulation of excess amyloid beta is the factor that sets off the series of events leading to the death of brain cells and is the main driver in causing Alzheimer's disease. The amyloid beta peptides* clump together to form larger particles called oligomers** and later these clump to form the plaques. One issue that has been plaguing scientists, however, is whether the amyloid causes AD or is a manifestation of the end process of the disease. Some researchers believe that it's the oligomers and not the plaques that set off the cell's destruction.

---

* Peptide- two or more amino acids linked in a chain
** Oligomer – a protein fragment or molecule that usually consists of amino acid residues linked in a polypeptide chain

It has been found that certain genetic mutations are associated with increased production of beta amyloid and the inherited forms of AD, which are relatively rare. On the other hand, plaques of beta amyloid can be seen with normal aging and those individuals affected show no signs of AD. Some feel that perhaps amyloid is still important but the location in the brain where it accumulates is the key factor in determining whether a person will show signs of Alzheimer's.

Other studies have been reported to show that the deposits of tau forming tangles actually correlate with AD symptoms better than the amyloid plaques. All this notwithstanding, significant efforts have been made to find a way to tie up, flush out, or in some other way get rid of the amyloid in the brain.

Drugs called bapineuzumab (Johnson & Johnson and Pfizer) and solanezumab (Eli Lilly & Co.) were developed and tested for their effectiveness in treating Alzheimer's disease.[16,17,18,19] The drugs were intended to help clear the amyloid beta protein plaques from the brains of people affected with AD. They are in the category of monoclonal antibodies and are a kind of immunotherapeutic agent. Basically they bind to the amyloid beta and are supposed to help flush it out of the brain and body.[17,18] Lilly's solanezumab, for example, helps to remove amyloid from the brain by attaching itself to the brain's free-floating amyloid before it clumps and forms plaques in the brain.

Bapineuzumab was reported to decrease nerve cell damage and the buildup of plaque in people genetically predisposed to Alzheimer's disease. In one study, scientists evaluated 1,100 individuals with mild to moderate AD who had a gene called APOE4 that puts them at risk for developing the disease.[17] Special imaging of those receiving bapineuzumab showed a significantly lower amount of amyloid in their brains and there were also lower levels of phosphor-tau, the other substance associated with AD, in their spinal fluid.[17] However, the new drug was not shown to affect brain shrinkage. In addition, there were some concerns about the drug's side effects. For example, fluid leaked into the brain from blood vessels in 15% of individuals, although most of them showed no symptoms. Seizures were also more common in the group of individuals receiving the drug, compared to a group of individuals getting a placebo.[17]

Another study evaluated two different dose levels of bapineuzumab in 1,331 individuals without the APOE4 gene.[17] There tended to be lower amounts of amyloid and phosphor tau but, again, there was an increased risk of fluid buildup and bleeding in the brain and of seizures. In addition, findings released in late 2012 revealed that bapineuzumab and solanezumab failed to improve memory in the patients studied.[17] Lilly reported that their drug, solanezumab did slow the mental decline in people with mild Alzheimer's disease but was not effective in the more advanced stages. In any case, it could not be said that either drug significantly improved the ability to think or improved the ability to perform daily activities in people with Alzheimer's disease.

Despite these disappointing findings, many researchers still believe that studies on amyloid beta should continue because of its prevalence in the brains of people with Alzheimer's disease. Some postulate that the drugs might be more effective in preventing AD rather than in treating it since the plaques are present in the brain years before a person is manifesting symptoms. The drug company Pfizer said it would not be doing further research on its drug bapineuzumab but Eli Lilly planned to do additional trials of solanezumab. Gantenerumab (Roche), another study drug and human monoclonal antibody, has also been examined for its potential to decrease beta amyloid by binding to amyloid clumps and facilitating their removal from the brain.[20]

Studies with bapineuzumab and solanezumab led to the conclusion that while the medications would not help patients who currently have Alzheimer's, there might be hope for their use in treatment of individuals who are very early in the course of the disease. Assuming these types of medications can be successful if used this way, the timing for optimal treatment results would have to be determined. There are, as always, financial implications as well. It has been said that drugs designed to slow the progression of Alzheimer's disease could create a huge market in sales, worth on the order of $20 billion dollars, according to some estimates. There would be an even larger market if trials are successful and show that the drugs work better if taken early in the course of the disease as more people will be taking the medications over a longer period of time.[17]

International efforts have also been undertaken testing different Alzheimer's drugs in their ability to prevent the disease in susceptible populations. One study involved 160 people from the United States, Britain and Australia with a variety of gene mutations known to cause Alzheimer's disease, with the majority of subjects not yet exhibiting any symptoms. In another investigation, an extended family from Columbia with the same gene mutation that causes AD was to be studied. The third study planned was that of asymptomatic Americans aged 70 years or above in whom imaging studies show early changes of the disease.[20]

Because of the high correlation that has been found between the presence of amyloid plaques in the brain and the development of Alzheimer's disease, there have been repeated attempts to find more effective drugs to diminish the amyloid. In addition to creating new drugs, medications used for other purposes have been targets of investigation. These other medications include the anti-cancer drug Bexarotene, the blood pressure drug Prazocin and a drug used for epilepsy called Levetiracetam,[21] The drug company, Biogen Idec, has experimented with a drug known as $BIIB_{037}$.[22] In an early stage of testing, the drug caused a decrease in the amyloid plaques and the subjects showed better cognition. Biogen planned further trials to study the drug's efficacy in early stage Alzheimer's patients.

In any case, no study up to now has resulted in a medication that has appreciably affected the course of Alzheimer's or the well being of patients, especially in the later stages of the disease. The studies have only held out a hope for a way to delay the processes of deterioration as early as possible before damage is irreversible. Still not definitively answered are the questions: Is amyloid the cause of Alzheimer's? Is a precursor of amyloid the culprit? Is it tau? Something else?

Why Has It Been So Hard to Find a Cure?

According to a study by the Cleveland Clinic Lou Ruvo Center for Brain Health, the failure rate for drugs developed to treat Alzheimer's disease was 99.6% during the period from 2002 to 2012.[23] Apparently, one drug showed some effectiveness while 244 others failed. The Clinicaltrials.gov website was the source of the data analyzed for this study. During the ten years, there were only 413 AD drug trials. The success rate for advancing from one clinical phase to the next was low and only an extremely small number made it to the regulatory review stage. The Ruvo Center director and lead author of the study, Dr. Jeffrey Cummings, noted that only eighty or so drugs were being tested to treat AD compared to about 300 drugs being tested to fight cancer.[24] The authors of the study believe that part of the problem is due to lack of funding for research.

In addition to the financial constraints, there may be a number of factors responsible for the failure to find a cure, or at least an effective treatment. A hypothesis increasingly put forth by researchers is that the medications are started when the disease has progressed too far and can no longer be reversed.[25] This has lead to a further impetus in finding ways to detect Alzheimer's disease as early as possible so that patients can begin taking the drugs earlier in the course of their disease. The hope of this avenue of research is to attack amyloid before it can do irreversible harm. Scientists now think that the disease process causing Alzheimer's actually begins about ten to twenty years before symptoms appear so it would take many years before there were meaningful results. However, while there have been many proponents of the drug trials targeting amyloid, the critics of this type of research have stated that, at best, results only *suggest* that targeting amyloid may be beneficial but there is no proof of this as yet.

Another possibility for the failure, therefore, is that the drugs are not targeting pathways which are the true key to development of Alzheimer's disease because these pathways have yet to be discovered. To date, the working theory has been that plaques of beta amyloid peptides and neurofibrillary tangles consisting of tau protein combine to block communication between neurons in the brain. However, the theory seems incomplete and the missing information may hold the key to discovering the "right" medication. How exactly does the action of these proteins lead to the actual malfunction and degeneration of the nerve cells? This question has stimulated other avenues of research.

To say that the working of the brain is extremely complex is an understatement! The task of finding a prevention or cure for Alzheimer's disease is daunting, since all the triggers and precise biochemical and neurologic changes that cause the clinical symptoms are not yet really known with any certainty.

So far, the finding of isolated bits of information which *might* lead to a cure have not been the successes they seem to be in news media accounts but they do hold out hope. It is possible that someone will stumble on a scientific breakthrough. Otherwise, and more likely, it would appear that it will take some time to build on all the new pieces of information to find a cure, and then to prove that it is actually effective. It is all worth the effort!

What Other Research Has Been Going On?

In addition to the various drug trials that have been initiated to find that prized medication that will either be the elusive cure, or at least a meaningful treatment to help those with AD live a much better life, many studies have been undertaken to try to elucidate the cause or combination of factors that will bring about the development of Alzheimer's disease in some individuals and not in others. The results of a number of these studies seemed promising at first but we no longer hear about them, others are still ongoing and many others are relatively new. Unless you are a neuroscientist (and I am not), reading about these can sometimes be bewildering as the information produced by the studies is not only technical but often preliminary, and, not infrequently, controversial. However, as we often hear or read about "new Alzheimer's research" in the media, it is enlightening and interesting to look at what scientists have actually been studying, where they have succeeded and what are the obstacles to progress.

Ginkgo Biloba: The Verdict is In

The herbal supplement ginkgo biloba has been popularized as a way to preserve memory and prevent Alzheimer's disease. Many studies have reported that it does not prevent or slow the progression of AD[26,27] A French study was one of the more recent of these and involved over 2,800 people, 70 years of age and older, who reported having problems with their memory. The subjects were assigned randomly to take either ginkgo biloba or a placebo. After five years, 4% of those taking ginkgo biloba developed Alzheimer's disease compared to 5% of those taking the placebo, a difference not considered significant.[27] There was also no real difference in the number of strokes or deaths between the two groups. However, it was noted that while the dementia rates were steady for the first four years in both of the groups, the people taking ginkgo did better in the last year of the study while there was a spike in onset of AD in the group taking the placebo. While the overall conclusion was that ginkgo biloba does not have a preventive effect against AD, this observation raises the question of how long should trials of new medications go on to be sure that a late effect is not missed and highlights the complexities of the research in this area.

Intravenous Immunoglobulin

Some investigators began working with intravenous immunoglobulin (IVIG), a component of plasma from human blood and results of a second round of testing were presented at the Alzheimer's Association International Conference in July 2012. This involved a study of only 24 people over a three year period. It was postulated that IVIG contains antibodies to the most toxic form of amyloid and will neutralize the toxins. There were some mild infusion related reactions, such as rashes, in the initial study. One person in the trial had a stroke but overall the drug appeared promising in slowing the rate of shrinkage of the brain and in slowing cognitive decline. However, results of a phase 3 randomized trial with IVIG were disappointing. There were 390 patients at 45 centers in the United States and Canada. The subjects received either IVIG or a placebo every two weeks for eighteen months. There was no significant difference in ability to function or in the rate of cognitive decline between the groups.[28,29]

Key Enzymes

As research has extended beyond examining amyloid plaques and the neurofibrillary tangles of tau, the language of Alzheimer's disease has expanded as well. News accounts now speak of **APP**, or **amyloid precursor proteins**, that are the origin of the amyloid plaques. The APP is cut or cleaved by certain enzymes* into fragments. The enzymes involved are alpha-secretase, beta-secretase and gamma-secretase. Depending upon which enzyme is involved and where the APP is cut can lead to different outcomes. When beta-secretase first cuts the APP molecule it ultimately leads to beta amyloid peptide being released into the space outside the

---

* Enzyme- a substance that speeds up a chemical reaction but which is not changed or used up in the process

nerve cell and the pathway ending in AD. Some researchers have been studying a class of drugs called BACE inhibitors.[30] These drugs block beta-secretase. The goal is to prevent the development of Alzheimer's very early in the process.[31]

Brain Pacemakers

A whole different direction of research involves implanting pacemaker like devices in the brain to provide deep brain stimulation (DBS) in patients with early stage Alzheimer's disease. Implanting brain electrodes has been successfully used in patients with Parkinson's disease and other neurological movement disorders with few side effects. Globally, 85,000-100,000 people have been treated this way. In fact, some investigators also began studying whether stimulating parts of the brain could help patients with depression or help curb appetite in the obese.

In 2003, researchers in Canada brought back memories in a patient they were treating for obesity. The DBS treatment also improved his ability to learn and the researchers wondered if this could be applied to patients with dementia. A team of scientists used the implanting technique on six Alzheimer's patients in Canada. Following twelve or more months of continuous stimulation, brain scans of the study subjects showed more glucose utilization and more activity in regions of the brain usually targeted by AD.[32] There was no decline in cognition in one Canadian who had the implants for four years although there is no way to know if that was due to the treatment. The Canadian researchers also partnered with researchers at four medical centers in the United States. In half of the patients involved in these studies, the DBS would turn on two weeks after the surgery and the other half would have the DBS turned on in a year to see if any changes were due to placebo effect shortly after the surgery.

The theory behind the DBS type of therapy has been that the electrical stimulation might keep the neural circuits involved in memory intact longer. It would not change the basic destructive changes of AD but hopefully bypass them. However, according to a recent study, six patients with Alzheimer's disease who received DBS in an area of the brain called the fornix demonstrated a mean decrease in hippocampal atrophy over a year compared to control subjects with AD,[33] suggesting that this type of treatment might be able to affect the structural changes of AD as well. Of course, DBS is more invasive than medications, since holes are drilled into the patient's head and the tiny wires must be properly implanted in the correct area of the brain. In addition, there is no way to know how long any beneficial effects will last or even if the treatment will be successful at all, and if there will be any adverse effects. This avenue of research certainly does highlight the frustrations and disappointments in finding effective drug therapy!

Looking at Changes in Brain Activity and Blood Flow

A study called the Baltimore Longitudinal Study of Aging looked at changes in brain activity over a period of time, beginning when the subjects of the study were cognitively normal. These individuals were evaluated yearly with cognitive tests and special types of positron emission tomography (PET) scans to measure blood flow in the brain. They were studied for periods of up to 17 years, with an average study period of twelve years. The earliest symptoms of cognitive impairment appeared at around the eleventh year. Of 121 individuals studied, 99 people were found to have maintained normal cognition while the 22 other individuals became significantly impaired cognitively. There were changes in brain activity patterns in the aging patients who remained cognitively normal but larger changes were seen before recognizable symptoms in the people who went on to develop cognitive impairment. There were larger increases in blood flow in centers for memory, attention and executive function in the brain's frontal lobe. These individuals also had greater decreases in brain activity in the parietal lobe in an area associated with memory and attention and in still another area associated with visual memory in the temporal lobe. Most of the changes in brain activity were found in areas most vulnerable to the accumulation of amyloid and tau. Changes in blood flow, therefore, seemed to reflect the underlying disease process of Alzheimer's.[34]

Cynthia Janus, M.D.

Leaking Blood Vessels in the Brain and Resealing the Blood-Brain Barrier

Researchers at the University of Southern California have found that the blood-brain barrier* can become leaky as a person gets older. These changes were found to begin in the hippocampus, that part of the brain that has been the focus of many studies since it is an important center for memory and learning and is affected by Alzheimer's disease.[35] Possibly a way could be found to reseal the leaky blood-brain barrier to protect the brain from the toxic effects of substances in the blood.

Using Ultrasound and Opening the Blood-Brain Barrier

Scientists in Australia used mice bred with a genetic defect that causes AD. They treated half of the mice with ultrasound to the brain. The treated mice did better on behavioral tests. For mice, this refers to things like the ability to navigate a maze. Post mortem studies demonstrated that the area of the brain's cortex occupied by plaques was 56% smaller when compared to the untreated mice.[36] The researchers believe that ultrasound may help by temporarily opening the blood-brain barrier. This would allow albumin, a type of protein, to enter the brain, helping cells called microglia to remove toxic proteins. However, skeptics feel that there are already drugs that clear amyloid so ultrasound may not be any better. In addition, it is difficult to test this theory on people for a variety of reasons, including the difficulty in penetrating the human skull with ultrasound waves.

Studying Mice and Toxic ADDL's

Research in mice has focused on a substance called amyloid beta-derived diffusible ligands (ADDLs), which are very small, soluble amyloid aggregates, discovered by William Klein, PhD at Northwestern University in Evanston, Illinois in the 1990's.[34] These are toxic forms of amyloid beta that attack synapses (the region of contact between two nerve cells.)

---

* Blood-brain barrier – a highly selective permeability barrier that separates the circulating blood from the extracellular fluid in the central nervous system; prevents many substances from leaving the blood and entering the brain tissues

Many researchers feel that the ADDL's are what cause Alzheimer's disease and not the amyloid plaques which have been associated with late stage AD. Klein's group has developed a probe with an antibody that binds to the toxic ADDLs and when fused to magnetic nanoparticles, they can be detected by an MRI scan.[34] The findings could lead to ways of targeting Alzheimer's disease for diagnosis and treatment in its earlier stages.

Studying Tau in Mice

In addition to the finding of beta amyloid between nerve cells, you will recall that twisted tau protein threads have also been found in the brains of Alzheimer's patients, especially in areas associated with memory function. Although great emphasis has been placed on amyloid, many researchers now believe that the tau may also play an important role in affecting the sequence of events leading to the development of Alzheimer's disease. Studies in mice have shown that tau frees brain cells from toxic proteins such as beta amyloid.[37] Without the beneficial influence of tau, the amyloid builds up and forms plaques between the brain cells which leads to a greater chance of brain cell death. The failure of tau to function properly could be the result of a genetic abnormality, the aging process or other factors. There has been continued and increased interest in the role of tau with some researchers speculating that abnormal tau protein is indeed a key factor, and perhaps even *the* key factor in causing Alzheimer's disease.[38] If proven true, drugs that promote or target tau in some way could be developed to fight Alzheimer's. Already, early in 2015, it was reported that Johnson & Johnson made a deal potentially worth over $500 million with a Swiss Biotech company, AC Immune, to develop vaccines against abnormal tau.[39]

Studying the Immune System and Arginine Consumption in Mice

Using a mouse model of Alzheimer's disease, researchers at Duke University have found that microglia, the immune system cells which have a protective effect in the brain, started to consume a nutrient called arginine.[40,41] Scientists used a drug called difluoromethylornithine (DFMO) to block this process in mice. They were able to halt the damage caused by the enzyme arginase which breaks down the arginine. Thus, we see another, at least theoretical, possibility of preventing memory loss in people.

Identifying Changes in a Gene for Earlier Diagnosis

The genetic aspect of Alzheimer's disease has also been researched. In the setting of the **early onset** type of AD, an association has been found with mutations on chromosome 21. They cause formation of the substance called amyloid precursor protein or APP, thereby affecting the production of amyloid. A connection has been noted in people with Down Syndrome who have three copies of chromosome 21 instead of two copies. There is known to be a high incidence of Alzheimer's disease in people who have Down Syndrome. The Alzheimer's symptoms can manifest very early, before age fifty.[42]

Likewise, mutations on chromosome 14 and chromosome 1 are associated with abnormality of the genes presenelin 1 and presenelin 2[43] respectively. These mutations are believed to influence the splitting of the abnormal precursor protein. In ways yet to be determined, the bad types of beta amyloid and eventually the amyloid plaques associated with AD are produced. Mutations of these genes are also associated with **early onset** Alzheimer's disease.

The genetic aspects of the much more common **late onset** type of Alzheimer's disease have also been studied. Currently, a gene called the apolipoprotein E, or **APOE** gene, located on chromosome 19 is the strongest genetic risk for AD after age 65. The presence of the four allelic variant of APOE is not only a risk factor for late onset Alzheimer's but may also be associated with an earlier onset of memory loss and more amyloid plaques in those affected. It had actually been implicated as a risk factor in studies going back to the 1990's.[44]

There are three different forms of alleles* of the APOE gene. The least common type is the APOE e2 which appears to reduce risk and is found in 5-10% of the population. The most common form, occurring in 70-80% of the population, is the APOE e3 allele which has not been found to confer either an increased or decreased risk of AD. The form with

---

* Allele - one of several forms of a gene; usually refers to one member of a pair of genes that control the same trait and that occupy a specific location on a chromosome

significant negative implications is the APOE e4 allele carried by 10-15% of individuals. The presence of this form of the allele increases a person's risk of developing Alzheimer's disease by three to eight times or more, depending on whether the individual has one or two copies of the allele. Not surprisingly, it has been shown that those people with the e4 allele are likelier to develop senile plaques of which the primary proteinaceous component is the infamous amyloid beta protein. Note, however, that some people with an APOE e4 allele will never develop Alzheimer's disease and many individuals who are diagnosed with Alzheimer's do not have any copies of this allele.

In addition to the correlation with increased amyloid plaques, some research has also shown a connection between APOE e4 and a significance decrease in a protein called SirT1 which is associated with anti-inflammatory effects.[45] Could finding a way to increase SirT1 help people with Alzheimer's?

Until about 2009, APOE e4 was the only gene known to be a risk factor for Alzheimer's disease. Scientists have been finding other genes that may cause an increased risk of developing AD as well as looking for genes which might have a protective effect against the disease. As part of the Alzheimer's Project Act of 2011, a genome sequencing effort was funded. Researchers have been using whole genome sequencing to detect genetic variations in the entire complement of a person's DNA.

In 2012, the New England Journal of Medicine published a report announcing the discovery of a new gene that almost triples the risk of AD and which also powerfully affects the immune system. Under normal conditions, the glial cells or microglia, the white blood cells in the brain, help to prevent inflammation and get rid of the toxic, plaque forming beta amyloid protein. However, when a gene called TREM2 is mutated, the white blood cells are not able to effectively attack the beta amyloid. The plaque is not removed and there are inflammatory effects. Indeed, researchers in Iceland found a variant of TREM2 more often in patients with AD than individuals in a control group of people without the disease. Their findings were repeated and confirmed at other institutions in the United States, Germany, Norway and the Netherlands.[46]

The TREM2 variant gene appears to have a role in the development of AD that is as strong as the APOEe4 gene. People who have the mutated gene are said to be three to five times more likely to develop Alzheimer's disease. However, genetic mutations in this gene are rare and having it does not ensure that you will develop Alzheimer's. Upon reviewing genetic data from around the world, the mutation was found in one-half to one percent of the general population. This increased from one to two percent in people with AD.

Researchers at the Icahn School of Medicine at Mount Sinai in New York City identified disturbed gene networks in areas of the brain that were most damaged by AD.[47] The network that was most affected had genes that controlled the brain's immune system. The glial cells, which function to remove cellular debris (including beta amyloid plaque) and infectious agents, were particularly affected.

Other researchers have searched for gene sequence variations associated with elevated levels of the biomarkers tau and phosporylated tau (p-tau) in the cerebrospinal fluid and four sets of DNA sequence variations were found. Two of these were located in APOE and TREM genes and the other two were in other locations.

Scientists at Massachusetts General discovered that higher levels of a substance called CD33 protein are produced by microglia in the brains of individuals with Alzheimer's disease when compared to people without AD. It has also been found that microglia in the brains of mice without the CD33 gene were more effective in clearing out beta amyloid. In addition, a CD33 gene variant in human brains decreased the formation of CD33 and amyloid, thus showing a protective effect against AD. This sounds promising but another "however" is in order. This is because other researchers found a CD33 gene variant that causes increased CD33 levels and more beta amyloid which would be associated with an increased risk of AD. Although somewhat confusing, these results contribute additional data points to the pool of information and are considered promising.[47]

Reversing Memory Loss in Early Alzheimer's Disease with a Drug for Epilepsy

Subjects in this research had what is called amnestic mild cognitive impairment (aMCI). People with mild cognitive impairment may have problems with memory, reasoning and other functions of the brain but they can still function quite well independently. MCI has been considered, however, as a significant risk factor or precursor of Alzheimer's disease and it has been described as a middle ground between age related cognitive decline and dementia. MCI is being studied more intensively and has been broken down into further categories. Memory loss, disproportionate in degree for the person's age, is the key presentation in amnestic type MCI. Researchers at Johns Hopkins found that very low doses of an atypical antiepileptic drug called levetiracetam reduces hyperactivity in the hippocampus, which is a feature of amnestic mild cognitive impairment.[48] This was demonstrated by means of a functional MRI exam. The subjects given the drug also performed better on memory performance evaluation.

A Drug for People With Moderate Stage Alzheimer's Disease

A drug called ORM-12741 has been studied with the hope of keeping memory loss at bay in patients with moderate Alzheimer's disease. A small Finnish study over a three month period looked at 100 patients who were already taking a standard cholinesterase inhibitor drug for AD. Some of the patients were also on the drug called memantine. Fifty of the 100 patients were given the new drug in addition to their regular medicine for the disease while the other half were given a placebo. The fifty patients who received the study drug showed a 4% increase in memory test scores while the memory scores dropped by 33% in the 50 patients who received the placebo.[49] The research was funded by Orion, the maker of ORM 12741. The drug is said to be the first to target a specific receptor in the brain, called alpha-2C. This receptor is thought to play a role in the brain's "fight or flight" response. It was considered promising that positive results were shown in only three months. However, initial research results always need to be duplicated, especially when performed with a larger study population. Generally, most experts believe that it is too late to even try to treat patients who have moderately severe symptoms of Alzheimer's.

Another Look at Vitamin E and Memantine

An investigation was undertaken to see if vitamin E (alpha tocopherol), the drug memantine, or both slow the progression of mild to moderate Alzheimer's disease in a group of older veterans taking an acetylcholinesterase inhibitor.[50]

The study, which ran from 2007 to 2012, involved 613 patients with mild to moderate AD who were taking an acetylcholinesterase inhibitor (donepezil, 65%, galantamine, 32%, or rivastigmine, 3%) They were divided into four groups. One group got synthetic vitamin E (alpha tocopherol, 2,000 IU/d), another group received memantine (20mg/d), while a third group was treated with both. The researchers found that treatment with 2000 IU/d of Vitamin E resulted in slower functional decline during a mean followup time of 2.3 years compared to those taking a placebo. However, for those individuals taking memantine alone or memantine with vitamin E, there was no more improvement seen over those those taking the placebo.[50] Currently, memantine is approved by the FDA for use in moderate to severe Alzheimer's disease. It may be used in patients with milder Alzheimer's symptoms although evidence of its benefit in the earlier stages of AD is lacking. Further investigation would be necessary with large population groups taking different forms of tocopherol and different dosages. In this study, as has been the case with other investigational drugs, any modest beneficial effect on symptoms did not translate into an actual reversal of the disease process.

Cooperative Effort by Pharmaceutical Companies

In February, 2014 it was announced that ten large pharmaceutical companies would take part in a government backed effort to cooperate in order to speed up the development of new medications for treatment of a variety of disorders including Alzheimer's disease.[51] The collaboration is called the Accelerating Medicines Partnership and involves these drug companies and the National Institutes of Health (NIH) which will share data and scientists as well as blood and tissue samples.

**Looking for Biomarkers for Early
Diagnosis: Predicting the Future**

Much of the research currently being carried out is concentrated on finding changes in the body which would definitively pinpoint the diagnosis of Alzheimer's as early in the disease process as possible and scientists are looking for biomarkers. **Biomarkers** may be specific proteins in the blood or cerebrospinal fluid or specific findings on brain imaging.

Examples of these biomarkers are the beta amyloid deposits in the brain measured by PET scans, and levels of beta amyloid, tau and phosphorylated tau (the more toxic form of tau) found in the cerebrospinal fluid. These and other biomarkers may provide information about an individual's risk of developing Alzheimer's disease. They can also be used to measure the progression of the pathology involved in Alzheimer's and could be used to help determine a person's response to a proposed treatment.

Remember that prior to the stage of actual dementia, individuals go through a stage of mild cognitive impairment (MCI). People with mild cognitive impairment have memory loss which is greater than that expected for their age. However, those affected by MCI can still function independently and their daily life is not severely impacted. On the other hand, once a person is in the stage of actual dementia, there has been death of brain cells with atrophy and destruction of brain tissue. At this point, memory loss is more pronounced and their ability to perform routine tasks and activities is compromised in different ways and in varying degrees. In these later stages of dementia, the dysfunctional abnormalities and disabilities are irreversible.

Patients with memory loss as the main symptom of their mild cognitive impairment are said to have a 10-15% chance of moving into the category of Alzheimer's disease. Currently doctors monitor changes in cognition with simple clinical tests and interviews. If accurate biomarkers are identified and used appropriately, medications hopefully developed in the near future, would have the best chance of being most effective.[52]

Spinal fluid tests can be used to **confirm** a diagnosis of AD but some scientists have also been studying proteins in the cerebrospinal fluid which may be forecasters of Alzheimer's disease. That means that some day it might be possible to **predict** the development AD by analyzing the person's spinal fluid. In one study,[53] samples of cerebrospinal fluid were taken from 137 people with mild cognitive impairment. Measurements of amyloid and tau, considered biomarkers of AD, were made. Ten years later, 72 patients had AD and 21 had other types of dementia. Evaluation of the patients' baseline tests showed decreased levels of beta amyloid-42 (a form of beta amyloid) and increased levels of Total tau (T-tau) and Phospho-tau (P-tau) in the patients with AD compared to patients who did not develop AD. Those who developed AD within 5 years of the baseline test also had decreased levels of beta amyloid-42 but they did not have increased levels of T-tau and P-tau. The tau protein remained stable until relatively close to the time that dementia symptoms became noticeable but the beta amyloid protein levels changed a decade before the onset of dementia.[53] This study would appear to lend support to the theory that changes in the metabolism of amyloid precede the effects of tau and the eventual death of nerve cells.

Scientists have also been trying to find other biomarkers in the blood which would be used to accurately diagnose and monitor the course of Alzheimer's disease and perhaps even help explain how it develops. A substance called brain-derived neurotrophic factor (BDNF) was shown to affect survival and function of neurons and improve long term memory in animal studies.[54,55,56,57] Levels of BDNF decrease during the course of AD. Interestingly, levels of the substance were shown to increase with physical activity and were also affected by caloric intake.[55] More questions are raised: What is the relationship between levels of BDNF in the brain and in the circulatory system? Is there a direct correlation between the amount of BDNF and the development of AD or are levels of BDNF a reflection of physical activity and caloric restriction acting as an indirect indicator of risk for AD? Can BDNF be used as a biomarker for AD which can possibly be modified with lifestyle changes? Further studies are necessary to determine the significance, if any, of BDNF in predicting Alzheimer's disease.

Other researchers have reported on the development of a blood test that could one day determine whether a person will develop Alzheimer's disease based on lipid levels. To investigate the possibility of this new potential screening test, in one study blood was drawn from hundreds of healthy people over age 70 in New York and California. Five years later, 28 of the subjects developed mild cognitive impairment or Alzheimer's disease itself. It was found that this subset of 28 people had low levels of 10 lipids compared to the non-affected people. The scientists also found that blood tests of 54 other people with mild cognitive impairment or who had AD also showed low levels of these lipids. As the next step, the researchers planned to try out the test on a group of younger people in their 40's and 50's.[58]

Other Possibilities for Diagnosing Alzheimer's ASAP: Do They Pass the "Sniff Test"?

There are additional examples of how scientific research seems to have pivoted toward finding biomarkers and diagnosing Alzheimer's disease before symptoms are apparent. Recent studies report on the development of other new screening tests for AD which would be noninvasive, meaning that they would have no risk or little risk of harming the patient, and simple to perform.

In one study, a decrease in the sense of smell predicted which individuals would develop Alzheimer's disease.[59] In this trial, 215 elderly people with no memory loss were given a scratch-and-sniff type test. After that, they underwent extensive evaluation including annual cognitive assessments and genetic analysis. They had MRI scans of the brain to measure a part of the brain called the entorhinal cortex which is involved in processing smells and short term memories. They also underwent PET scans to evaluate for beta amyloid in the brain. A smaller hippocampus and thinner entorhinal cortex correlated with poorer memory and poorer results on the smell test. In subjects with elevated amyloid levels in their brains, there was also a positive correlation with thinner entorhinal cortex and poorer olfactory discrimination

According to other research, Alzheimer's disease could potentially be diagnosed before symptoms actually appear by the detection of the substance beta amyloid in the lens of the eye. In one study, individuals were given a substance that binds to beta amyloid. They underwent eye exams using a technique called retinal amyloid imaging and then had PET scans of the brain. Preliminary findings showed a correlation between beta amyloid in the eye and in the brain.[59]

There are, however, limitations to both studies. We already know that not everyone with beta amyloid in the brain (and presumably in the eye) will develop Alzheimer's. As to the scratch-and-sniff testing, it is known that many other conditions can adversely affect a person's sense of smell such as sinusitis, allergies and a history of smoking.

Researchers say that the exams could be used to screen people for further, more extensive testing, clinical trials, or earlier treatment. However, one should keep in mind that these seemingly innocent screening exams have the potential to be more harmful than good. First of all is the question of the impact of a positive result on individuals and their families. Just being told that you *may* have Alzheimer's disease is life altering and a frightening prospect. Should an individual even go down that road on the basis of a test which may be inaccurate? Another harmful consequence is the potential surge of pseudo- medical professionals offering questionable screening exams for AD and undoubtedly charging a large fee.

Focusing on Treatments for Preclinical Alzheimer's Disease

We have seen that in light of the failures of clinical trials testing potential drugs to treat Alzheimer's disease, attention has now turned in large part toward studying the earliest stages of the disease before symptoms are apparent. This stage, which has been called **preclinical AD,** may predate the occurrence of clinically evident dementia by ten or more years. At this point, it should be noted that the term preclinical Alzheimer's disease does not mean the same thing as early onset AD. In fact, as the pathologic mechanisms causing late onset AD may differ from people with early onset disease, many scientists recommend that current investigational studies should be performed in both groups of patients.

Traditionally, preclinical Alzheimer's disease has been determined to be present if there are positive findings on amyloid positron emission tomography (PET) scans or on analysis of the cerebrospinal fluid in people not manifesting signs of AD. However some medical and research personnel have come to believe that PET scans may not be sensitive to the earliest stages that occur before the accumulation of enough amyloid in the brain to show up on the scan.

Trying Out New and Better Imaging Techniques to Diagnose AD

Using an MRI imaging technique called diffusion tensor imaging (DTI), researchers have seen abnormalities in white matter connections of the brain in patients with certain types of Alzheimer's disease.[60,61] One study looked at 53 people, some of whom had Alzheimer's disease and others atypical types of AD affecting localized parts of the brain. AD has been considered a gray matter disease yet patients with atypical forms of AD and with early onset disease showed extensive damage to the white matter along with local areas of damage in the gray matter.[61] A theory is that AD may spread along white matter fibers from one area of the brain to another. It is hoped that the special imaging technique of DTI could be used as an imaging biomarker to identify the white matter degeneration earlier in the evolution of the disease in people before symptoms are apparent.

Another imaging development is the use of a technique called arterial spin labeling (ASL) MRI which measures brain perfusion or how well the blood is getting into the brain. The patient does not have to be injected with a contrast agent for the test. In one study, individuals underwent the MRI exam and neuropsychological testing at baseline and 18 months later. There were 148 asymptomatic people and 65 individuals with mild cognitive impairment. In the asymptomatic group, 75 people remained symptom free and 73 individuals showed some cognitive decline at 18 months. The people who ultimately declined showed reduced blood perfusion in an area of the brain called the posterior cingulate cortex on the baseline MRI. The pattern was similar to the individuals with mild cognitive impairment. These areas of decreased brain perfusion were in the same locations in the brain that have been known to show AD associated abnormalities on PET scan.[62] These findings may lead to a new screening imaging exam which, unlike the PET scan, does not expose the patient to ionizing radiation or require an injection of contrast.

Update on Amyloid

Two large studies from the Netherlands were published in the Journal of the American Medical Association in May 2015 and focus on the significance of beta amyloid as it relates to Alzheimer's disease. In one study, titled *Prevalence of Cerebral Amyloid Pathology in Persons without Dementia: A Meta- analysis*, it was found that an increased prevalence of amyloid pathology, shown on PET scans or in cerebrospinal fluid evaluation, correlated with age, APOE genotype and the presence of impaired cognition.[63] Subjects aged 18 to 100 years included 2,914 individuals with normal cognition as well as 697 with subjective cognitive impairment (SCI) and 3,972 with mild cognitive impairment (MCI). In SCI, an individual feels they have memory loss but cognitive tests are normal.

People having the APOEe4 allele demonstrated increased production of amyloid and an earlier age of onset of Alzheimer's disease when compared to people having the APOEe3 allele. The presence of the APOEe2 allele correlated with a lesser likelihood of finding amyloid and an onset of AD at a later age. There was also a relationship between education level and amyloid. When compared to people with lower educational levels, there was a higher prevalence of amyloid and a later onset of dementia in those who had more education. The presumption is that education may offer some protection against AD, possibly because the brain is handling and processing information in a different and more efficient manner. As far as people with mild cognitive impairment, there was found to be a 20-30% higher rate of amyloid than in individuals with normal cognition or subjective cognitive impairment. This supports the general assumption that having mild cognitive impairment is a risk for developing Alzheimer's. Not everyone with mild cognitive impairment had amyloid and it was postulated that the cause of their MCI might have been some other disorder. Another important finding of this research analysis was that beta amyloid in the brain could be present twenty to thirty years before symptoms of dementia are apparent. The authors concluded that we cannot know with certainty if, and when, a person without dementia but who has the amyloid findings will go on to develop dementia. The finding of amyloid does not guarantee that an individual will have dementia in his lifetime but he or she should be considered at risk.

The second study is called *Prevalence of Amyloid PET Positivity in Dementia Syndromes: A Meta-analysis.*[64] The subjects of the study included 1,359 people with the clinical diagnosis of Alzheimer's disease and 538 individuals with other types of dementia. Researchers found that in the individuals with dementia, the prevalence of amyloid was correlated with age, clinical diagnosis and APOE genotype. In this study, 88% of people with Alzheimer's disease demonstrated amyloid pathology while fewer people with the other dementias had amyloid. In those people with the diagnosis of Alzheimer's disease, the prevalence of amyloid on PET scans decreased with age while it increased with age in people with other types of dementias. Results of the study may help to determine at what age and under which circumstances imaging for amyloid could be most helpful in evaluating a person for Alzheimer's disease. Both studies may help provide information as to which groups of people are more at risk for developing AD and ultimately in helping to determine who might benefit from therapies directed against amyloid.

Summing It Up

Researchers have been looking everywhere to get a better understanding of what changes take place in the brain to cause a person to get Alzheimer's disease and to try to find better ways to detect it as early as possible. The end goal, of course, would be to develop a drug that would treat or cure it, or even better, prevent it. The large number of drug trials and scientific investigations should be promising but are considered insufficient by many. It might just take one study to find the key to developing the cure. To date, however, results have been disappointing,

Lack of sufficient funding is often blamed for inhibiting progress. Another reason for failure to find an effective medication may be that the areas of research have not been wide enough, with the vast majority of studies targeting the importance of amyloid plaques and perhaps even these studies were done when the patient's disease was too far advanced for medications to be efficacious.

Scientists are grappling with many questions at the same time. Is it really the presence of amyloid alone that causes Alzheimer's or is the amyloid just a by-product of the disease? Do other factors interfere with normal brain function independently, or in concert with the amyloid? Are some people more susceptible to getting AD than others? If so, why? Is there any way to prevent getting Alzheimer's disease? If not, can it be stopped early in its course? How long **is** its course? How early can we identify the anatomic changes of Alzheimer's in the brain and how can we track them? Often the research brings up more questions than answers.

III CONTROVERSY AND UNCERTAINTY

Amyloid, PET Scans and Screening

Trying to Define Alzheimer's Disease

In dealing with Alzheimer's disease on any level, as a patient, caregiver, or researcher, you soon find out that there are many unsettled issues and areas of controversy. For patients and their loved ones, there are often questions about treatments. Why does a particular medication seem to work better for one person than another? Why did it seem to be working but then stop? Why are scientists taking so long to find or produce the right drug?

For those involved in the science of AD, there have been differences in opinion over the years as to what causes the brain pathology. Beta amyloid has been the object of considerable investigation and discussion and continues to be considered a key component of the disease.[1,2] However, there has been a change in scientific opinion as to whether the amyloid is the actual cause of the dementia and other symptoms of Alzheimer's or just a late by-product of some other pathologic processes going on in the brain. Research has shown that finding a greater amount of amyloid plaque in the brain is associated with a higher risk of developing Alzheimer's disease yet a significant number of people found to have the amyloid plaques do not display signs of dementia.[3,4] One theory proposed to reconcile these confusing and opposing facts has been that buildup of the deposits leading to AD is such a long, slow process that these individuals just didn't develop the inevitable dementia by the time they died. It is also possible that even if the amyloid deposits are the inciting event for development of AD, there are other factors that influence the course of the disease. Still another idea is that perhaps the *location* of the amyloid plaques, rather than just the amount present, is the factor that determines whether Alzheimer's disease will develop in the individual.

Recently, there has been increasing emphasis on the role of the precursors of amyloid, the tau protein and other substances in bringing about Alzheimer's disease. Of course, there may be other undiscovered factors responsible for causing AD which are not yet being studied.

It is important to sort all this out for many reasons. It is difficult to come up with a cure for Alzheimer's without knowing exactly how the disease comes about. If the exact factors that cause AD were known, it would be easier to test for it and predict how severely it might affect a given individual. In addition, it would be helpful to have some specific criteria for the diagnosis of AD in order to determine which individuals are eligible to participate in a clinical treatment trial.

The Role of PET Scans

Many people have heard of PET (positron emission tomography) scans which have become increasingly important and extremely useful in the diagnosis and follow up of cancer and other diseases. They have also been used as part of the evaluation of individuals for Alzheimer's disease. As part of a PET scan, a radioactive substance called a tracer is injected into the patient's vein and is carried in the bloodstream to the organs. The tracer shows up on images from the scan when it binds to certain substances in areas of the body. In this way PET scans can give information about brain function or chemical makeup including the presence of abnormal substances.

There are two types of PET scans that have been used to assess dementia. In an FDG (fluorodeoxyglucose) PET scan, the degree of activity in a portion of the brain is reflected by the amount of glucose (sugar) utilized. Centers in the brain which are involved in memory, speech and higher reasoning have been associated with lower levels of glucose utilization in those people with Alzheimer's disease. The newer, second type of PET scan, or amyloid PET scan, was FDA approved in 2012 to find beta amyloid plaques in the brain. In conjunction with these scans, the radioactive tracer is injected into the patient's vein in order to evaluate for the presence and assess the extent of the amyloid plaques characteristic of AD.[3] Amyvid (florbetapir), produced by Eli Lilly & Co., was the first compound approved by the FDA for use with PET scans for this purpose.[3] Other companies, including General Electric (GE), Piramal Enterprises LTD and Navidea Biopharmaceuticals Inv (NAVB), also worked to develop tracers to be used with the PET scans.[3]

The use of PET scans, especially for screening purposes has, however, been a topic of controversy. Proponents of the use of amyloid PET scanning for diagnosis and screening believe that detection of amyloid plaques by the scan translates into earlier diagnosis and treatment of people with Alzheimer's disease and they will then benefit from a better quality of life. They believe that in order to make an accurate diagnosis of Alzheimer's and optimize treatment, there needs to be a way to confirm the presence of the amyloid plaques, since it is the hallmark pathologic change of the disease.[5]

In this regard, research has shown an association between the finding of amyloid plaques and the presence of, or seeming propensity to develop, Alzheimer's disease. In one study, a group 152 adults, fifty years of age and older, consisting of 69 individuals with normal cognition, 52 with mild cognitive impairment and 31 with Alzheimer's dementia, underwent PET scans and cognitive tests at the start of the study and again, 36 months later. Those people with no cognitive issues or with mild cognitive impairment but who were found to have plaques on their initial amyloid scans had worse results on followup cognitive tests than those people whose amyloid scans were negative for plaques. In addition, 35% of people with mild cognitive impairment and plaques went on to develop Alzheimer's disease compared to 10% of people without plaques. On the other hand,

90% of people with mild cognitive impairment but no plaques did not progress to AD.[6] Other research, including the large studies from Finland described in the last chapter also stress the significance of amyloid in Alzheimer's disease.

Those on the other side of the argument feel that the scan comes with a high price tag which may not be justified. The cost of amyloid PET scans varies but can be upwards of $3,000.[7] They call attention to the annoying fact that although considered to be a hallmark of the disease, not everyone with amyloid plaque has signs of Alzheimer's or will ever display symptoms in their lifetime. In fact, according to some estimates, 30% of people demonstrated to have amyloid plaques do not and will not show signs of AD.[7] More significantly, the use of scanning has not been shown to improve patient outcomes. This speaks to the most compelling argument of the PET scan skeptics that making an earlier diagnosis and being able to more accurately predict the course of the disease is all well and good, but what use is it really if you can't effectively stop or significantly delay and diminish the terrible symptoms of the disease for the patient?

Financial Implications

We know that current medications for Alzheimer's disease provide only temporary and variable symptomatic improvement at best but if drugs are ultimately developed which can successfully prevent or circumvent the damage to brain cells caused by the amyloid plaques, the use of PET scans will increase dramatically. This raises more concerns. Are the PET scans accurate enough to justify the cost? Who will pay?

On January 30, 2013, a Medicare advisory panel consisting of twelve members called the Medicare Evidence Development and Coverage Advisory Committee (MEDCAC) met in order to give guidance to the Centers for Medicare and Medicaid Services (CMS) as to whether the agency should pay for the brain imaging scans for amyloid detection. The panel gave a low to intermediate confidence level that there was sufficient evidence that imaging for amyloid plaque in the brain had any effect (good or bad) on people with early signs of cognitive dysfunction. On a scale of 1 for "low confidence" to 5 for "high confidence," the panel assigned a score of 2.1667.[8] The vote displayed a lack of certainty that these scans could improve the outcome of patients with early cognitive dysfunction. More importantly, perhaps, is the concern that the issues stemming from receiving the diagnosis of Alzheimer's disease with no hope of meaningful treatment or the potential negative effects of receiving a false positive result would far outweigh any benefits .

One could argue for the value of scanning for amyloid in that it may help in understanding the course of the disease, leading to the ultimate development of a cure. Also, it might help doctors rule *out* Alzheimer's disease as a cause of cognitive decline in a patient. Another justification for the use of the amyloid PET scan is that it is more quantitative than psychological testing for Alzheimer's. In addition, traditional neuropsychological evaluation is not considered adequate to differentiate AD from other types of dementia.

In terms of screening, however, there is also the question as to whether the scans have been around long enough to feel confident in making treatment decisions based solely on their findings. The Medicare advisory panelists noted that the scans for amyloid did not bring about any change in an individual's outcome or any change in the course of care by their physician. Eli Lilly and Company, the pharmaceutical company that produces the florbetapir F-18 (Amyvid) injection agent for the amyloid scan, was said to be disappointed that CMS denied patients access to its drug.[9]

The use of amyloid PET scanning was supported by the Society of Nuclear Medicine and Molecular Imaging and the Alzheimer's Association. Two days before the MEDCAC meeting, these organizations drew up guidelines for when the amyloid PET scans should be used.

According to the recommendations of these organizations, appropriate candidates for PET imaging include people with persistent or progressive unexplained memory problems or confusion and who show impairment on standard tests of memory and cognition, those testing positive for possible AD but who have an unusual clinical presentation and people with progressive dementia with an atypical age of onset, before 65 years of age.[3,10] People who are **not** considered appropriate candidates are individuals 65 years or older who already have satisfied the standard criteria of AD. Another group who were deemed inappropriate for amyloid PET imaging are those who complain of cognitive loss but there has been no clinical confirmation.[3]

Since the development of Amyvid, other new agents have been approved to be used in the evaluation of brain amyloid on PET scans in people being investigated for the presence of cognitive impairment or dementia. In addition, new PET imaging agents have been developed to detect abnormal tau.

Advancing the Discussion on Amyloid or Going in Circles?

After all the controversy regarding the value of reimbursement for amyloid scans, the Centers for Medicare and Medicaid Services (CMS) has approved a $100 million, four year study called Imaging Dementia – Evidence for Amyloid Scanning (IDEAS). The study will evaluate how amyloid imaging performs in clinics in the "real world" away from the realm of academic medical centers and clinical trials.[11] It is supposed to help determine whether undergoing such a scan will affect diagnosis, treatment and outcomes in people in whom a definite diagnosis cannot be made by their clinical picture alone.

IDEAS is a huge study which will analyze approximately 18,500 amyloid scans performed at 200 or so imaging centers across the country. The costs of the scans will be covered by the CMS under its coverage with evidence (CED) application. CMS will pay for one amyloid scan per person taking part in a clinical trial if data is collected about how the scan affected the person's outcome. There are specific criteria for people with mild cognitive impairment or dementia to be included in the study and also specific criteria that the referring doctors have to meet.

The subjects in the study will be scanned using one of three FDA approved tracers: florbetapir (Avid/Lilly), flutemetamol (GE Healthcare) or florbetaben (Piramal). The referring doctor fills out a form with the patient's diagnosis and treatment plan before the scan and then sends in a second report three months after the scan with a new diagnosis and treatment plan. The physicians will be reimbursed for the time spent in filling out the paperwork.

The study will examine if the amyloid scans resulted in a change in the patient's diagnosis and/or treatment plan and determine if it reduced the amount of other, unnecessary testing. The researchers also hope to see if the scans decreased the number of hospitalizations and emergency room visits by reviewing Medicare claims for the following year.[11] In a prior pharmaceutical company study using florbetapir (Amyvid), the management plan of 68% of patients examined was changed, particularly by the addition of cholinesterase inhibitor drugs, if scan results were available to the physician immediately when compared to 56% of subjects who received their results one year later. However, there was no difference in cognitive performance or quality of life at one year.[12] The Metabolic Cerebral Imaging in Incipient Dementia Study, sponsored by CMS showed that patients with early Alzheimer's disease receive medications earlier and have better outcomes after two years when the patients' doctors were given PET scan results right away.[13]

The IDEAS study is likely to help those people who turn out to have negative amyloid scans in that the doctor can look for other, possibly treatable, causes of their symptoms.[14] While the presence of beta amyloid plaques confirms but does not guarantee that the patient has Alzheimer's, the absence of the plaques is felt to pretty much exclude the diagnosis. If this is true, these patients would also likely not be mistakenly entered into Alzheimer's trials.

Those people with positive amyloid scans may or may not have Alzheimer's disease.[3] If they are already showing signs of cognitive impairment and they have a positive amyloid scan, present day medications are only of limited and temporary help. In addition, if the patients are cognitively impaired and treatable conditions have been ruled out, lifestyle changes and financial planning would have to be implemented whether they had a scan or not. These issues and the incomplete fund of knowledge underlying them lead to the question that follows.

Would You Want to Know???

Due to the recent attention given to Alzheimer's disease, many older people start to worry if they are in the early stages if they misplace their keys or wallet, can't remember a name, or forget why they just walked into a room. If they express this concern out loud, a friend or loved one will usually tell them reassuringly that it is only "a senior moment". Has this happened to you? Now, if a blood test, imaging exam or other screening test was developed that was accurate in predicting the development of Alzheimer's disease, WOULD YOU WANT TO KNOW?

Some people would **not** want to know that they were destined to endure a progressively debilitating disease for which there is no cure. On the other hand, others might want to prepare themselves and their relatives for the inevitable and plan for the future as best they could

There are several factors to consider. The screening exam used should be precise, accurate and well tested. Most sources say that cognitive test results can be affected by many factors such as educational level, sensory deficit and psychological state. The "jury is out" on the accuracy of the amyloid scan. In addition, to date, there is no proof that early diagnosis improves patient outcomes. Many scientists, however, believe that Alzheimer's disease is not caught early enough for proposed treatments to be effective which would speak to the need for screening to detect the early cases.[15, 16, 17]

It is also reasonable to assume that early diagnosis of Alzheimer's disease could be beneficial in that it would enable the patient and his or her family to make appropriate plans. The individual would also have time to make sure that their finances will be protected. Safety measures, such as evaluating the need to stop driving, could be instituted.

At the same time, people contemplating aggressive screening for Alzheimer's disease should be aware of the tremendous consequences this diagnosis will bring, including the potential for significantly impacting life plans and financial decisions, as well as the ability to qualify for long term care insurance. Since there are no effective treatments, getting a diagnosis of Alzheimer's disease will only lead to an earlier recognition of the usually frightening road ahead and bring about a loss of any sense of well being, if not extreme anxiety

Whether you would "want to know" or not, consider this: you may not be given the choice! According to the Alzheimer's Association, surveys have shown that a significant number of people with Alzheimer's disease and caregivers of people with AD were never informed of the diagnosis by their physicians.[17] You might think that some people with AD might not have understood or forgotten what their physicians told them or that they might be in denial. However, only about half of caregivers claimed to have been given the diagnosis. Could it be that physicians are uncomfortable telling patients that they have Alzheimer's when they have no effective treatment to offer them? It takes time and compassion for doctors to explain the ramifications of the disease and answer questions. According to the report, much higher percentages of patients with cancer or heart disease say that they were told of their condition by their doctor. This is a different situation from screening individuals before the disease manifests itself. People with symptomatic Alzheimer's disease and their caregivers have the right to know the diagnosis and should be given the information and guidance that they will need.

In addition to individual personal concerns and preferences regarding screening, other factors come into play. One question is whether all, or any, new screening techniques will be covered by insurance. Medicare has paid for diagnostic evaluations for Alzheimer's disease but only if cognitive impairment has already been found. In 2011, Medicare funded an annual wellness visit which was to include a cognitive assessment to look for early signs of AD. The Affordable Care Act requires doctors to check Medicare patients for cognitive impairment during the annual wellness visits. Many applaud this new change, including the Alzheimer's Association, believing that those people with previously undiagnosed cognitive disorders will now gain access to medical care.

Skeptics, however, cite that a large part of the screening for AD has traditionally involved family members filling out a questionnaire about their loved ones. There is no specific standard test or guidance for follow up. Also to be considered is the point made by some that early dementia is difficult to distinguish from those familiar "senior moments" and other signs of mild cognitive decline we experience as we get older. Depression, alcohol abuse and vascular dementia can also be falsely diagnosed as Alzheimer's disease because of the similarities in some of the symptoms. The quick type of screening done during a regular office visit may, therefore, be inadequate.

The screening test must be a good one because think of the consequences of getting an incorrect diagnosis of Alzheimer's disease. Overdiagnosis caused by inappropriate or inaccurate screening could cause people to be referred for unnecessary further testing.[18] They may suffer side effects of medication that they don't need, not to mention the considerable negative psychological effects this diagnosis will bring to them and their loved ones. According to a study published in the British Medical Journal online in 2013, up to 23% of adults without dementia could be misdiagnosed by early screening.[15]

Therefore, in addition to the need for guidelines as to who should be screened and when, there must also be standards to ensure that the screening test has some validity. For example, according to media reports, the FDA found that two UCLA researchers were promoting an unapproved drug called FDDNP for their company, Taumark. Patients are injected with the FDDNP for a scan performed to detect concussions and some types of dementia at an early stage. The scans were supposed to show tangled protein deposits of tau in individuals with CTE (chronic traumatic encephalopathy). CTE is a neurodegenerative disorder with symptoms of memory impairment, loss of impulse control and other symptoms, some of which are similar to those of Alzheimer's disease. The term refers to a condition which was seen in boxers (formerly called the "punch drunk" syndrome) and which has more recently been seen in football players. Outside experts maintain that the value of the test has not been proven and that autopsies have shown that the same protein deposits have been found in people without symptoms.[19] The controversy illustrates the point that any screening method has to be thoroughly tested and this generally involves large groups of subjects and a long period of time to know if the findings and predictions are accurate.

The last issue to consider is that of age. At this time, age itself is not considered an indication for screening for dementia. The United States Preventive Services Task Force is an independent panel of experts that makes recommendations regarding screening and other preventive services. The panel has declared that currently there is not enough evidence to support screening of all older adults for dementia or for mild cognitive impairment.[20] Critics say the findings of the task force are based on cost-effectiveness analysis. In other words, if there is no cure for dementia, why spend money to screen for it?

However, while widespread screening of the elderly is currently considered unnecessary, or at least debatable, the presence of symptoms should not be overlooked. Loved ones should take note if an individual is forgetting the names of family members or common objects, displays inability to follow a conversation, misplaces things in strange places or seems disoriented. These symptoms are a cause for concern and evaluation of the individual would be most important. Many experts feel that the best and first way to make a diagnosis is examination by a medical specialist, particularly a neurologist or gerontologist who has expertise in this area. A thorough assessment including a full evaluation of memory, judgment and orientation to time and place would be necessary to evaluate for the presence of Alzheimer's disease or other type dementia and refer the individual for appropriate further testing.

Summing It Up

Scientists have been looking for ways to screen people for Alzheimer's disease even before there are any symptoms. They hope to be able to diagnose it as early as possible so that medications to prevent or treat symptoms would have the best chance of working. PET scans have been the most publicized of the screening tests but other methods of screening are also being proposed. Of course, a clear understanding of how Alzheimer's disease develops is necessary to develop an accurate screening exam and this is currently not possible. Another great concern involves the ethical and psychological issues related to screening for AD.

IV DO WE KNOW WHY SOME PEOPLE ARE MORE VULNERABLE?

Is There a Way to Avoid Getting Alzheimer's?

We have established that scientists do not yet know the precise mechanisms in the brain that will lead to the development of Alzheimer's disease. There is no effective cure or prevention and there is no medication to adequately treat, certainly over the long term, the symptoms of Alzheimer's disease. Do we at least know what makes some people more vulnerable than others? Are there risk factors or risky behavior that we can avoid to lessen our chances of developing AD? Numerous factors have been examined but in many cases the studies have raised even more questions. Initial research findings into what might make some individuals more susceptible to AD have been debunked in some cases and the studies have been discontinued while research in other areas is ongoing.

Cholesterol

Elevated cholesterol is associated with a higher risk for heart attacks and strokes and has also been considered as a predictor of dementia. Past research has shown that, at autopsy, there were higher levels of amyloid deposits in the brains of people with higher cholesterol levels.[1]

In a more recent study published in the journal JAMA Neurology, researchers from UC Berkely, UC Davis and USC, evaluated PET imaging of the brains of 74 older patients. They did not have dementia but over half of them showed mild cognitive impairment and they all had been treated for high cholesterol. Those with higher levels of LDL cholesterol (the "bad" cholesterol) had more amyloid plaque deposits in areas associated with Alzheimer's disease. Those with low levels of HDL cholesterol (the "good" cholesterol) also tended to have more amyloid plaque. No correlation was shown between triglyceride levels and total cholesterol and the extent of amyloid plaques. Approximately one-fourth of the cholesterol in our body is located in our brain but it is separated from the cholesterol circulating throughout the rest of our body by the blood-brain barrier so it is not known how the apparent association between cholesterol levels and amyloid plaque could arise.[1] Again, more questions than answers.

If there is a true association between elevated cholesterol and buildup of amyloid plaques leading to Alzheimer's disease, can statin drugs, which are commonly prescribed for people with abnormal lipid profiles, also be used to prevent dementia? Findings here are a bit confusing. In 2012, the FDA added a warning to the label on statin medications because short term studies showed a higher risk of memory loss and confusion in patients on these medications. However, more recently, researchers at Johns Hopkins University found no adverse cognitive effects for people using statins for less than a year. In addition, longer term use of these drugs may reduce the risk of dementia. It is postulated that statins, which suppress cholesterol formation in the liver, also cut down formation of plaques in blood vessels which would otherwise compromise blood flow to the brain and adversely affect cognition. In addition, the ability of statins to reduce inflammation may be beneficial to blood vessels and that would help also maintain better cognition.[2]

High Blood Pressure, Silent Infarcts and Diabetes

Studies have looked at any correlation between amyloid plaques, Alzheimer's disease and small silent infarcts (microinfarcts or small strokes) in the brain caused by disease of the small blood vessels.

One such study included 914 individuals and ran from January 2002 to December 2012. Findings showed evidence of greater deposition of amyloid in patients with disease of the small blood vessels in the brain, especially in those who carried the gene for APOE e4 which is associated with Alzheimer's.[3]

In another study, autopsy results on 774 Japanese American men who were subjects in a study called the Honolulu Asia Aging Study showed that those men who had been treated with medication for hypertension (high blood pressure) had fewer microinfarcts as well as fewer amyloid plaques and tangles and also less atrophy in their brains. Is there a connection between these microinfarcts which indicate small strokes and the amyloid plaques and tangles associated with signs of Alzheimer's disease? The brains of those men who had been on beta blockers for their high blood pressure showed the least damage.[4] Other research, however, has failed to show that aggressive treatment of blood pressure helps to ward off AD.[5]

The origin of some of the research on the effect of treatment regimens for high blood pressure and diabetes is the ACCORD (Action to Control Cardiovascular Risk in Diabetes) trial. The subjects of this trial consisted of 10,251 middle-aged and older individuals with type 2 diabetes. Researchers found that the subjects undergoing more aggressive treatment to reduce their blood sugar levels did no better as far as cognitive loss than those getting standard treatment. In addition, this portion of the trial was discontinued in 2008 because there was an increase in death rates with more aggressive treatment of blood sugar levels.

Approximately 2,977 people in the ACCORD trial also participated in the MIND (Memory in Diabetes) trial which focused on intensive versus standard treatment of high blood pressure and dyslipidemia*. As for blood pressure, there was an attempt to bring down the systolic reading to 140 with standard treatment and to 120 in those treated more aggressively. For standard measurement of lipids, a statin drug was used to lower levels of LDL cholesterol whereas those people treated more aggressively also received an additional medication to lower triglycerides and raise levels of HDL cholesterol.

---

* Dyslipidemia – refers to increased total or LDL cholesterol levels or to decreased levels of HDL. The abnormal levels are considered to be risk factors for coronary heart disease and stroke.

It was found that the use of medications to aggressively reduce blood pressure and lipid levels was no better than standard therapy at decreasing cognitive loss. Furthermore, aggressively lowering systolic blood pressure appeared to be associated with more brain shrinkage. As part of this research, 503 individuals had MRI exams of the brain at the beginning and at the end of a 40 week study. There was a significant decrease in the volume of the brain of those whose high blood pressure was more aggressively treated than in those who received standard treatment. A conclusion drawn by the researchers is that earlier prevention works better to ward off cognitive loss and dementia than aggressive treatment later on.[5]

Insulin Resistance in the Brain

It is believed that insulin plays an important role in normal brain function. At optimal levels, it has a protective effect on memory but problems with cognition can occur if insulin levels are too low or too high or if the brain becomes resistant to its beneficial effects. Insulin also helps to regulate the levels of amyloid beta. Studies have shown an association between low levels of insulin and cognitive impairment. Individuals with type 2 diabetes have a higher risk of developing Alzheimer's disease than people with normal glucose levels.[6] Some researchers have considered the possibility of developing medicines that would help to resensitize the brain to insulin or finding a way in which insulin could be delivered directly to the brain.

Eating Too Much

Researchers have been planning studies in humans to see if intermittent fasting can reduce the risk of Alzheimer's disease. It is based on what has been felt to be encouraging findings in mice in studies performed by the U.S. National Institute of Aging. There was no reduction in brain plaques and tangles in mice. Instead, there was suggestion of increased production of a protein called brain derived neurotrophic factor or BDNF that aids in memory and learning.[7] Interestingly, some people have independently been fasting two or more days a week in an attempt to prevent a wide range of disorders including obesity, cancer, and dementia.[7]

Midlife Stress

Swedish researchers have found that a great deal of stress in middle age is associated with a higher risk of dementia later in life. They reviewed the records of 800 women covering a period of approximately 40 years. The subjects were part of a larger Prospective Population Study of Women which began in 1968. The women underwent neuropsychiatric exams and evaluations when they entered the study and again in later years. They were asked about the psychological impact of various life stressors including divorce, widowhood, death of a child, unemployment and so on. Between 1968 and 2006, 153 (19%) of the women developed dementia and 104 of these were diagnosed with Alzheimer's disease. It took 29 years, on average, for dementia to develop and the diagnosis was made at the average age of 78 years. There was found to be a correlation between the number of stressors reported and the risk for developing Alzheimer's. Women who were more prone to anxiety and psychological distress were at greater risk.[8]

Emotional stress is known to affect physiologic reactions in the central nervous system as well as the cardiovascular, endocrine and immune systems. It has also been shown by other studies that stress hormones can remain at high levels many years after a traumatic event. What role stress plays, if any, in the development of AD, however, has yet to be defined.

Apathy

At the other end of the emotional spectrum, the presence of apathy has caught the attention of some. One study looked at approximately 4,300 people, mostly in their 70's, from Iceland. Older adults showing signs of apathy tended to demonstrate a smaller brain volume on MRI exam, with slight decrease in both gray matter and white matter. The findings raise the question of whether apathy is an indicator of increased risk for dementia associated with Alzheimer's disease or with other brain disorders. It is known that people with dementia, including Alzheimer's disease, often show apathy, but none of the people being studied had dementia. In any case, the researchers warn of jumping to any conclusions since the study doesn't prove which came first: the apathy or the brain changes.[9]

Depression and Cognitive Decline

Some studies have associated the presence of depression with a tendency toward more rapid cognitive decline. This is very controversial, however, and some experts feel that the depression and the cognitive decline may both be a reflection of some other underlying mechanism. Researchers at the Rush Alzheimer's Disease Center in Chicago examined the records of 1,700 individuals with no cognitive loss who were over the age of 50 years. During a period of eight years, they were evaluated for cognitive loss and for depression on an annual basis. Approximately half of the group developed mild cognitive impairment and 18% developed dementia. Those individuals who showed more symptoms of depression on their early exams were more likely to develop dementia and also had a more rapid decline. According to the study, the symptoms of depression decreased once dementia was diagnosed.[10] This is surprising. It seems intuitive that receiving a diagnosis of dementia and then experiencing the social isolation that usually goes with it would increase depression. Another study in Germany involved 371 patients with mild cognitive impairment (MCI). Those individuals with MCI and symptoms of depression were shown to have more amyloid deposition in the brain on scans than those people who were not depressed.[11] Sometimes, doctors prescribe anti-depressants for people with dementia. However, there is no evidence that these types of medication will affect progression of the cognitive decline.[10,12]

Living in a Country with Better Hygiene

Living in a wealthy country with better access to clean water and lower rates of infectious disease and a higher percentage of the population living in urban areas sounds like a good thing, right? Maybe not so much. Researchers analyzed data from the World Health Organization's (WHO) Global Burden of Disease (GBD) report in 2009 and found that there were higher rates of Alzheimer's disease in these areas according to their study published in the journal Evolution, Medicine and Public Health. According to the "hygiene hypothesis," people who live in areas with better hygiene have less exposure to certain germs resulting in insufficient development of their immune system, making them more at risk for developing certain allergies and autoimmune diseases.[13] Can Alzheimer's disease be added to the spectrum of these diseases or do the statistics merely reflect under-reporting of Alzheimer's disease in developing countries?

Head Injury

A study of veterans by the University of California-San Francisco and the San Francisco Veterans Affairs Medical Center found a link between brain injury in veterans and a higher risk of developing Alzheimer's. They examined the records of 188,764 veterans, 55 years of age and older, who did not have dementia at the beginning of the period surveyed. They found that 16% of vets who had severe head injury were diagnosed with dementia by the end of their study period, compared to only 10% of the vets with no history of head injury. The presence of other conditions including post-traumatic stress disorder, heart disease and depression led to a further increased risk of developing dementia.[14] Does a brain injury lead directly to the development of amyloid plaques or in some way make the brain more vulnerable to existing plaques? The scientists point out that we cannot conclude that brain injury leads to AD, as the brain injuries of the vets were of a much more severe nature that those that might occur in the general population.

The condition called chronic traumatic encephalopathy (CTE) has also been associated with a higher incidence of developing dementia although some studies have yielded inconclusive results. Many of the symptoms are similar to those seen in Alzheimer's disease but occur in a much younger age group. CTE has been associated particularly with those individuals engaging in football and boxing. Deposits of neurofibrillary tangles of tau protein have been found in the brain postmortem. More recently, there have been attempts to demonstrate these changes in living individuals by using PET scans.[14] The possibility of accurately screening for CTE and other dementias is also under scrutiny, as discussed in the previous chapter.

After a Stay in the ICU

An admission to the ICU is indicative of one or more serious medical problems. With skilled physicians and advanced medical technology, these medical issues can most often be successfully addressed. However, it has been noted that many of the patients suffer substantial cognitive deficits upon discharge. While this has been long known by ICU staff, it was recently quantified in a study by physicians at Vanderbilt University and Thomas Hospital in Nashville and published in the New England Journal of Medicine. The group studied consisted of 821 patients who were placed in the ICU for a wide variety of reasons including preceding trauma, surgery, downturn in a medical illness or a stroke. It was shown that three months after hospital discharge, four out of 10 patients had persistent cognitive defects, similar to what is seen in moderate traumatic brain injury and more than one-fourth had cognitive defects similar to those seen in patients with mild Alzheimer's disease.

The cognition deficits were still present one year after hospital discharge in 58% of the patients who had been in the ICU. Not only the elderly were affected; the sudden cognitive decline was also found in young patients. The length of time a patient experienced delirium* appeared to be the only factor that the researchers uncovered which correlated with which of the patients ended up with cognitive problems. There was a weak correlation between those given benzodiazepines, including diazepam (Valium). There was otherwise no clear correlation between the painkillers and sedatives widely used in the ICU. The researchers also postulated that the delirium itself is associated with inflammatory changes and the death of brain cells which may initiate or signal a downward spiral in brain function.[15]

Not Enough Sleep

A study on mice, funded by the National Institute of Neurological Disorders and Stroke (NINDS), has led to the theory that sleep has the potential to restore brain function by facilitating the removal of certain molecules that are linked with the process of neurodegeneration.[16] The need for sleep has persisted in humans and other living beings throughout evolution, so it seems reasonable to assume that it serves an important or useful function. One is advised to "get a good night's sleep" before an exam or a job interview and most have us have experienced difficulty in learning

---

* Delirium- the term used to to describe a disturbance of the mind and cognition that usually occurs acutely in association with fever, intoxication, and other medical disorders. Marked restlessness, agitation and incoherent speech are some of the symptoms.

and slower reaction times when we are sleep deprived. The reasons for these phenomena have not been completely elucidated.

In a study at the University of Rochester Medical Center, scientists evaluated changes in the fluid in the brains of living mice using special imaging. There was a 60% increase in the interstitial space* in the brain during sleep or anesthesia and greater exchange of cerebrospinal fluid with interstitial fluid.[16] The mice appeared to have greater clearance of amyloid beta while asleep. This study follows work published in the journal Science Translational Medicine by another research team which demonstrated that there is a sort of plumbing system known as the glymphatic system, located along the blood vessels, for getting rid of molecular debris in the mouse brain. It pumps cerebrospinal fluid through the brain tissue and gets rid of waste into the circulatory system where it is eventually dealt with by the liver.[17,18] Other researchers have attempted to link poor or disrupted sleep with the finding of increased brain amyloid on imaging scans or with deficits shown on memory tests.[18] Whether this translates into poor sleep being a prognostic sign for the development of Alzheimer's has not yet been proven, however.

A related investigation of the effects of poor sleep in people involved a cross-sectional study of 70 older adults living in the community. The information was obtained from a neuroimaging sub-study of a normative aging study (the Baltimore Longitudinal Study of Aging). The subjects ranged from 53 to 91 years of age with a mean age of 76. PET scans were used to measure the amount of amyloid, also referred to as the *beta amyloid burden*. After accounting for a variety of other factors, the researchers found that those people who reported poor sleep quality and fewer hours of sleep were found to have a larger beta amyloid burden.[18] The findings are interesting but additional studies with more objective measurements of sleep duration and quality would be important to further explore this issue. Can it be that poor sleep actually causes or speeds up the progression of Alzheimer's disease?

---

* Interstitial space - the fluid-filled spaces that surround the cells of tissues.

High Estrogen Levels Plus Diabetes

According to French researchers, older women with high estrogen levels may have a higher risk for developing dementia. The risk is further increased if they also have diabetes. From a pool of over 5,600 postmenopausal women, 65 years or older, the researchers measured estrogen levels in those women who did not have dementia and who were not on hormone replacement therapy. After four years, the researchers compared the baseline estrogen levels of 543 women from the study who did not have dementia with 132 women who had been diagnosed with dementia. The women with high estrogen levels showed over twice the risk for dementia even after other risk factors for dementia were taken into account.[19] The risk increased even more for those women with high estrogen levels who also had diabetes.

Estrogen levels generally decrease after menopause. However, some women, for example, those with more body fat, may have higher estrogen levels. Whether endogenous or by prescription, estrogens have been said to be beneficial for brain, as well as heart health but these teachings have been countered by the French study as well as other research. Some studies have shown that higher estrogen levels before age 65 decreases the risk for dementia, but in the elderly, higher estrogen levels appear to have a negative impact.[19,20] It should be noted that although the French study shows an apparent association between estrogen levels and dementia, a direct cause and effect relationship has not been defined. It also did not study the effects of hormonal therapy.

Being Female

The increased vulnerability to Alzheimer's disease by just being female has been of interest for a long time but is being studied more intensely of late. It is known that, of the more than five million people in the United States who have been diagnosed as having Alzheimer's disease, approximately two-thirds are women. Some have thought that this is because getting older is the major risk factor for AD and that women tend to live longer than men. The World Health Organization reports that, in the United States, the average life expectancy is 76 years for men and 81 years for women. According to results of the Einstein Aging Study at the

Albert Einstein College of Medicine in New York, women who are 70 to 79 years of age are twice as likely to develop dementia, including Alzheimer's disease as men in this age group. The risk is similar after age eighty.[21]

Several possible explanations have been proposed. One is that men are more likely to develop and die from cardiovascular disease before they develop AD. The survivors might have a healthier cardiovascular status which might be protective against dementia. Another theory is that women have a higher rate of depression and some studies have uncovered links between late life depression and dementia. Still another possibility is that the risk for developing dementia may be greater in those with less education and this may be the case for older women whose academic and professional opportunities were more limited than they are today.

Many researchers have attempted to discover the possible role of estrogen in either bringing about or helping to prevent the development of dementia. Some have thought it might help delay or prevent Alzheimer's disease, possibly because of its anti-inflammatory effects. Other studies of synthetic estrogen, with or without progestin, seemed to show a greater risk of developing AD and other types of dementia.

Later efforts have studied the timing of the estrogen therapy. Some studies have shown that women given estrogen replacement therapy within five years of menopause seemed to have a lesser chance of developing AD. Women on estrogen therapy beyond five years did not show a decreased risk. There was an increased risk of AD for women taking estrogen with or without progestin late in life.[21]

Since women make up the majority of Alzheimer's patients, the role of estrogen seems worthy of further investigation. To this end, it is noted that female Alzheimer's researchers have advocated for more gender specific research.

Exposure to Some Common Infections

Preliminary results of some studies have indicated a possible link between certain bacteria and viruses and later cognitive decline, according to researchers from the University of Miami and Columbia University in New York City. Cognitive tests were performed on 588 participants with an average age of 71 years. The scientists also looked for signs of the C. pneumoniae and H. pylori bacteria and of the cytomegalovirus and herpes simplex viruses 1 and 2 by measuring antibody levels in the blood. Approximately half of the subjects had cognitive tests again five years later. Those with increased antibody levels, indicating prior infection or exposure to the bacteria or virus, did more poorly on the cognitive assessments. It had previously been shown that those individuals demonstrating the highest levels of infection or exposure had more plaque in their carotid arteries and a higher incidence of stroke. The authors postulated a link between vascular disease and loss of cognitive skills through pathways involving the immune system.[22]

Exposure to DDT

An article published in JAMA Neurology in early 2014, based on a study done at the Rutgers Robert Wood Johnson Medical School in New Jersey, links DDT exposure with later development of Alzheimer's disease. Scientists measured levels of DDE (dichlorodiphenyldichloroethylene) in the serum in 86 patients with Alzheimer's disease and in 79 control patients. DDE is obtained from the breakdown of DDT (dicholorodiphenyltricholoroethane) in the body. They found that blood levels of DDE were almost four times higher in people who had Alzheimer's disease when compared to those individuals who did not have the disease. Increase in levels of amyloid precursor protein levels by DDT and DDE may account for this association. DDT was banned by the Environmental Protection Agency (EPA) in 1972 because there were concerns of a risk to wildlife but the chemical had been used widely in the United States since the 1940's to kill the mosquitoes that cause malaria. DDT is still used in some countries, with approval of the practice by the World Health Organization in 2006. Humans can become exposed by eating contaminated fish, meat or dairy products. The DDT can persist in human blood and human tissues, as well as in the environment. The study did not go as far as to claim that DDT exposure can cause Alzheimer's. However, those people in the study group who performed the worst on memory and reasoning exams had two risk factors. They had a variation in the APOE gene, known to increase the risk of developing AD, as well as high levels of DDT exposure.[23]

Drugs Used to Help People Control Anxiety and Sleep Better

French and Canadian researchers have studied the effects of medicines called benzodiazepines. Some of the trade names of these drugs include Xanax, Valium, Ativan and Klonopin. Usage of benzodiazepine was compared in 1,796 people with Alzheimer's and 7,184 individuals without the diagnosis. Those people who took a low dose of a benzodiazepine drug or who took the medication at higher dosages but only infrequently or for a brief period of time did not have a higher risk of AD five years after they had started the medication. However, at the end of five years, the risk of development of AD was 32% higher for those who took the cumulative equivalent of daily doses for three to six months over the five year period and 84% higher for those who took the cumulative equivalent of a full dose for more than six months.[24]

Some researchers feel that the brain receptors to which these drugs bind become less active which can lead to cognitive decline in the individual who is using the medications on a regular basis. Medical guidelines already warn that benzodiazapines should not be used regularly for more than three months but many patients continue taking them for years.

Recent news reports have also cited research that seemed to show that the use of drugs that have anticholinergic effects might also cause an increased risk of dementia. The study involved following 3,434 adults, 65 years and up, for 17 years. None had signs of dementia at the start of the study but 797 people had developed it by the study's end. Those people who took a minimum of 10 mgs per day of Sinequan (the tricyclic antidepressant doxepin), 4 mgs per day of Benadryl (the antihistamine diphenhydramine) or 5 mgs per day of Ditropan (the anticholinergic bladder control medicine) for a period longer than three years had a greater risk of developing dementia.[25] The anticholinergic properties of some drugs block the chemical acetylcholine. This is the chemical transmitter in the nervous system which is deficient in people with Alzheimer's disease.[25,26] In animals, the anticholinergic effects of these drugs have been linked to an increase in beta amyloid.

Vitamin D Deficiency

Researchers at the University of Exeter Medical School have reported in the journal Neurology that there is a significantly increased risk of dementia in patients with moderate or severe vitamin D deficiency. The risk of dementia was 53% higher for those with a moderate deficiency and 125% higher in those with severe deficiency of the vitamin. The moderately vitamin D deficient individuals were 69% more likely to develop Alzheimer's type dementia and those severely vitamin D deficient were 122% more likely to develop AD. Vitamin D is known to be important in maintaining strong healthy bones but other effects throughout the body are also being studied. Some postulate that it may help regulate calcium levels in brain cells and/or have a role in eliminating beta amyloid plaques.[27]

A Specific Protein and Its Role in the Failure of the Brain's Stress Response System

Research by scientists at Harvard focuses on a protein named REST which may protect the neurons or nerve cells from stresses related to aging in older people who are healthy. There is significant loss of this protein in portions of the brain in the counterpart of those with dementia, including Alzheimer's disease. REST acts like a controller. The scientists discovered that there was a small amount of the REST protein in people from 20 to 35 years of age. The levels of REST increased with age in those without dementia. In individuals with mild cognitive impairment, AD and other types of dementia, there was much less of the REST protein in the areas of the brain affected in these disorders. There was also an interesting distinction between the people who actually had memory loss and cognitive dysfunction and those people whose brains showed the same amount of amyloid plaques and tangles of tau but who were asymptomatic for cognitive problems. It turns out that the asymptomatic group with the brain changes had three times as much REST than those showing the symptoms of mild cognitive loss or dementia. As symptoms worsened, the levels of REST fell, but only in the areas of the brain that involved memory, learning, planning and other functions necessary for optimal cognition.[28, 29, 30]

Why does this happen? This protein, REST is known to switch off certain genes in fetal neurons. When a baby is born and the brain begins to function, the REST is no longer active. What the researchers found was that REST turns out to be a very active regulator of genes in the brains of older people as well. A theory put forth by the scientists is that the brain is subjected to considerable stress both in the birth and aging processes, endangering neurons that don't have the capability of regenerating. Apparently, the REST protein can turn off genes that are conducive to the death of neurons.[28, 29, 30]

The results of this study are fascinating and could lead to an explanation of why the approximately one-third of people who have the Alzheimer's associated findings of plaques and tangles in the brain do not develop symptoms of the disorder. Could the cause of AD be related to some type of failure of the brain's complex system of responding to stress?

As with other discoveries, much more research is necessary to draw any hard conclusions. One limiting factor in studying this protein has been that there is no way to measure REST in the living person. Death itself may cause damage to brain cells making it difficult to know by evaluation of the brain that a decrease in REST caused the dementia.[28]

In a related study, mice that were genetically engineered to be deficient in REST lost neurons in areas of the brain associated with Alzheimer's disease as they aged. A working theory is that when the brain is stressed, REST goes to the nucleus of a neuron. In cases of dementia, REST goes to another part of the nerve cell with other toxic proteins where it is destroyed. At this point in time, the overall significance of REST in the etiology of AD is unknown.

Your Genes

Is Alzheimer's disease "in your genes"? There have been studies to examine the role of the APOE gene and more recently the TREM2 variant gene in Alzheimer's.[31] Scientists have actually discovered about 21 genes that increase the risk of AD. Eleven of the genes were relatively recently identified. DNA was analyzed from more than 74,000 people of European ancestry in 15 countries and the results were published in the journal Nature Genetics.[32,33] The study, which was the work of the Cardiff School of Medicine and the International Genomics Project, was praised as a significant contribution to the fund of information about Alzheimer's. Some of the new genes are associated with the immune response in the brain suggesting a role for the immune system in the development of Alzheimer's disease. It also enforces evidence of the role of certain genes in the buildup of amyloid protein in the brain.

On the other hand, other researchers have noted that, to date, only the APOE gene has a significant effect on increasing the risk of Alzheimer's disease, especially in people who inherit two copies of the variation.[32] Most of the other genes increase the risk of development of AD by only a very small amount, Some scientists using MRI techniques have been investigating a possible role of APOE in the normal development of the human brain. The goal is to determine the extent and nature of the effects of APOE in the development of AD later in life. Can therapies be targeted against the specific adverse effects of the gene?

APOE4 and Being Female

The APOE gene appears to affect men and women differently. According to a publication in the Annals of Neurology, the gene had only a minimal effect on men while it almost doubled the risk of getting mild cognitive impairment or Alzheimer's disease in women.[21,34] These results could be factored into the finding that nearly two-thirds of the five million or so Americans currently diagnosed with AD are women. The research brings up the question of the role of estrogen levels in developing memory impairment. Do women's brains become more vulnerable to whatever causes AD when estrogen levels are declining around the time of menopause? This recent line of scientific investigation also raises the issue of how to find a treatment or cure for Alzheimer's disease if the biologic processes that cause it to occur turn out to be different for men and women.

The Proteins Associated With APOE4

The question of why the APOE e4 variant increases a person's risk of getting Alzheimer's disease has been a focus of study at the Buck Institute for Research on Aging based in Novato, California. According to their research, the status of APOE e4 and the clearance of amyloid beta plaque in the brain may not totally explain the development of AD. They studied cell cultures of APOE e4 and determined that APOE e4 was associated with significant decrease in a protein associated with anti-inflammation and anti-aging properties. This protein is called SirT1 and, as it is lowered in amount, it affects the amyloid precursor protein or APP which has an important effect on the storage or loss of memories. The researchers identified four drugs that seemed to be able to maintain levels of SirT1 in APOE e4 cell cultures. The drugs have not yet been tested in humans.[35,36]

In addition, in some experiments, scientists were able to put SirT1 proteins back into cells that were already affected by APOE e4 and found that the abnormalities in the cell were corrected, If this could translate to humans, it would mean that treatments could be developed for people already in the early stages of AD.[36]

Family History

A family history of Alzheimer's is a known risk factor for development of the late-onset, more common form of the disease. In fact, the risk is two to four times higher for people who have had a mother, father, brother or sister with the disease.[37] First degree relatives share approximately one-half of their genes with another member of the family. Genetic variations, including those affecting the APOE gene account for about 50% but the other causes or genetic factors accounting for the risk relationship are not known.[37]

One study from the NYU School of Medicine examined 52 people between 32 and 72 years of age, none of whom had dementia. Those people whose mother and father both had Alzheimer's disease were found to have 5 to 10% more amyloid plaques than those individuals whose parents were not affected by Alzheimer's. Interestingly, in the cases where only one parent had Alzheimer's, there was more plaque in those people whose mother had AD when compared to those people whose father had the disease.[38]

The genetic component or influence on the development of Alzheimer's disease is not straightforward. In general, some disorders, including late onset Alzheimer's disease are not inherited by the simple laws of Mendelian inheritance that we learned in biology class. Instead, of a variation in a single gene, small contributions from variations in many different genes acting together, perhaps along with environmental influences, may be responsible for the familial occurrence of the disease.

Having Multiple Chronic Health Conditions

A study published in the Journal of the American Geriatric Society looked at 2,176 people, randomly selected from the community, who were cognitively normal. Of this group of people, 87% had two or more chronic health conditions (multimorbidity). Upon following these people for up to about seven years, it was determined that the risk of developing mild cognitive impairment or dementia was greater in those people with two or more chronic conditions compared to those with no chronic conditions. The risk was greater for those with four or more chronic conditions compared to those with two or three ailments and was higher for men than women. The authors concluded that there may be multiple contributing factors causing mild cognitive impairment and dementia in later life. [39]

Things That Might Protect You Against Alzheimer's Disease

While investigating genetic and environmental factors that may induce the onset, or make one more vulnerable to, Alzheimer's disease, some are also looking at things that may have a protective effect against the disease. This has resulted in studies by respected researchers as well as unsubstantiated claims by those who may merely be opportunists targeting the vulnerable. There have been many puzzle books, diets and supplements that have been marketed as a way of preventing Alzheimer's. While the makers and producers of these items stand to make a profit, the benefit to those people who buy them in an effort to try to improve their memory and prevent getting Alzheimer's disease is often questionable at best.

Supplements, Snake Oils and False Promises

Taking a food supplement to improve your memory and perhaps even prevent Alzheimer's disease sounds very appealing. Proving that these supplements actually work is another matter. As a case in point, Quincy Bioscience is the manufacturer of Prevagen products which are based on a synthetic jellyfish protein called apoaequorin. If you ever checked out their website, you will have read that "Prevagen improves memory in a 90 day study" and that "a majority of study participants reported Prevagen noticeably improved practical areas of memory like remembering driving directions and recalling words in a conversation in 90 days."[40]

Quincy and its claims have not gone unchallenged however. In October 2012, the company was sent a warning letter by the FDA charging that Quincy was selling unapproved new drugs rather than dietary supplements. The company's website had boasted claims of cures of memory disorders, including Alzheimer's, and the company failed to report over 1,000 incidents where people had bad reactions, some requiring hospitalization. Quincy removed the unsubstantiated claims from its website and retrained employees about regulations relating to supplements. However, the company denied that the bad reactions were necessarily due to the supplement.[41,42] The difference in terminology as to whether a substance is a "supplement" or a "drug" has important ramifications because there is much less regulatory oversight over a supplement when compared to a drug. Approval of a new medication by the FDA generally takes years and there has to be extensive clinical trials and testing.

In commercials for Prevagen, you might have heard, "Can a protein originally found in a jellyfish actually improve memory? Our scientists say yes!" Well, a class action lawsuit against the maker, Quincy Bioscience, in January 2015 said "NO." The class action suit said that the protein, apoaequorin, is destroyed by the digestive system and changed into common amino acids just like those that come from other food products. In addition, the studies mentioned in Prevagen's ads were said to be so flawed that they prove nothing and that claims that these clinical tests show that Prevagen results in better memory and sharper thinking are not true.[42,43]

Meanwhile, Prevagen is available at 20,000 drug stores including large chains like Walgreens and CVS and it is also sold online.[41] You can draw your own conclusions as to whether these or other supplements can help your memory but there is certainly no evidence to prove that they can prevent Alzheimer's disease.

The Brain's Ability to Compensate for Damage

Many researchers have been trying to understand why a significant number of people with amyloid buildup in the brain **never** develop

Alzheimer's disease. Results of a study published in Nature Neuroscience in September 2014 suggest that there may be mechanisms by which the aging brain can compensate for the destructive effects of the beta amyloid. A group of 22 young adults and 49 older individuals who had no symptoms of cognitive loss underwent functional MRI exams of the brain while they memorized various pictures. The subjects were questioned on details of the scenes in the pictures that they memorized. There were no significant differences in the performance results among the participants. However, there was more brain activity in the individuals who had beta amyloid deposits and suggestion that their brains were compensating for the amyloid by some yet unknown mechanisms.[44,45] Along these lines, other studies have suggested that people who have engaged in brain stimulating activities professionally or recreationally are potentially better able to adapt to the damage caused by the abnormal proteins.

Lifestyle Changes and Cognition.

While there are still news media reports of new scientific findings in Alzheimer's research, there also seems to be increased interest and reporting on various lifestyle modifications which might improve cognition.[44-50] At this point, it is important to talk about a concept called **cognitive aging** which refers to changes in cognitive function that take place as we get older. These changes are gradual and also very variable. For example, a person's reaction time may increase and memory may get worse even while overall knowledge may increase. Cognitive aging is **not** considered a disease. Studies of this process in animals have revealed that neurons do not die like they do in Alzheimer's disease but instead there is functional decline at the synapses, which are the contact points between the nerve cells.[51] The Institute of Medicine released a report called Cognitive Aging: Progress in Understanding and Opportunities for Action which included recommendations for ways to prevent or slow down cognitive loss by making lifestyle changes. Individuals with high blood pressure or diabetes, for example, can try to better manage these conditions and smokers should make every effort to stop smoking. Many believe that getting enough sleep, engaging in intellectual activities and staying socially active may also help cognition.

Apart from the Institute of Medicine report, investigators have been looking for ways in which people might be able to "ward off" Alzheimer's disease itself. In one study, the results of which were published in the American Journal of Preventive Medicine, brain MRI scans of 260 people were correlated with their diet. Those people who ate baked or broiled fish weekly were found to have more gray matter on the scans. The study implies a potential for better cognition or memory in these patients. However, no association was found between blood levels of omega-3 fatty acids (which are found in fish) and the amount of gray matter in these individuals.[52] Past reports on the effects of omega-3 fatty acids and retention of good memory have been conflicting.[53] Others have advocated the DASH (Dietary Approach to Stop Hypertension) diet and the MIND (Mediterranean-DASH Intervention for Neurodegenerative Delay) diet for cognitive health.[54] While these types of diets have been advocated for people with heart disease and diabetes, nutritional changes have only recently been touted as potential modifiers of Alzheimer's disease.

Another recent study goes further than the recommendation of specified foods by concentrating on a substance called resveratrol in high dose form. Resveratrol is an antioxidant that is found in red wine, red grapes, berries and dark chocolate. In the study, 119 adults with mild to moderate Alzheimer's and on current Alzheimer treatment drugs were given either resveratrol capsules or a placebo. Amyloid beta levels in the spinal fluid decreased in amount in those people who were given the placebo which is what normally happens when Alzheimer's disease progresses. However, in the group of people taking resveratrol, the spinal fluid amyloid levels stabilized suggesting that resveratrol might be able to slow the progression of AD. However, the study did not address the question of whether AD symptoms also stabilized in the people getting the drug.[55]

While some are researching the relationship of nutrition and cognitive decline, others are investigating the effect of a **combination** of lifestyle changes on decreasing the risk of memory loss. These changes include eating a healthy diet, exercising, managing blood pressure, obesity and diabetes and keeping mentally fit and socially active. In one study, a group 1,260 individuals, ranging in age from 60 to 77, from Finland were divided in half. One group was encouraged to practice the healthy lifestyle changes above while the other group had what was called standard care. After two years, the group which emphasized the healthy lifestyle changes did better on cognitive and memory tests. Both groups were to be studied for another seven years to see if the trend persisted.[56]

Still another interesting study looked into the effects of cataract surgery on people with dementia. It turns out that 28 patients who had the surgery declined much more slowly than 14 other patients who did not undergo the surgery.[57] That seems right. Why would depriving people with dementia of optimal eyesight be any good?

The studies above, along with many others,[58,59,60] are focusing on the potential to reduce the risk of Alzheimer's disease by altering various dietary, medical and psychosocial factors. Proponents of this strategy believe that individuals should be evaluated for diabetes, hypertension and other cardiovascular conditions and treated when indicated. They should be encouraged to exercise, lose weight and stop smoking to promote better health, which now includes brain health. Along with this line of thinking, there should be better access to education globally. Studies have shown an inverse relationship between level of education and risk for development of AD,[61] although a cause and effect has not been proven.

Unfortunately, however, while there is evidence that making certain lifestyle modifications can influence cognitive ability as we age and the changes generally have a positive effect on well-being, there is no evidence to suggest that they can prevent the onset of Alzheimer's disease.

Summing it Up

Researchers in the United States and all over the world have looked at just about everything and have uncovered a number of culprits that may be associated with an increased risk for developing Alzheimer's disease including genetic and familial influences and a variety of environmental factors. Unfortunately, no direct cause-and-effect relationship has been proven. Possibly, it takes the interplay of two or more of these, or other factors, to cause the disease. It is interesting that the vast amount of research being undertaken to find a pharmaceutical cure notwithstanding, emphasis seems to be returning somewhat to health factors in our environment that we can actually control. So while we wait for the discovery of the optimal biomarkers and drug treatments, it doesn't hurt to live a healthy lifestyle!

V LONG TERM CARE AND NURSING HOMES

Last Desperate Measures

People with Alzheimer's disease survive, on average, eight to ten years after the diagnosis is made[1] although life expectancy has been said to range anywhere from three to twenty years.[1,2] Ultimately, those affected will require some type of long term care to provide a safe place to live and help with personal care, such as eating, bathing and dressing. They also require attention to their individual medical needs, some of which may be complex. Most of these individuals will spend the end of their life in a nursing home, which has traditionally been the main symbol of long term care. Currently, 50-70% of the elderly in nursing homes have some degree of dementia.[3] People with dementia are more than seven times more likely to move into a care facility than those with other chronic illnesses such as cancer and heart disease. Approximately 70% of people with dementia die in nursing homes.[4] This contrasts with only 20% of patients with cancer and 28% of patients dying from all other conditions.

The growth and development of nursing homes has been shaped by financial rewards balanced by regulations brought about as a reaction to various prior abuses and scandals. Nursing homes have evolved to administer two different types of care and the expected outcome of each type is different. In one case the goal is to provide care, after the acute period, for people who have been discharged from the hospital. This is

Medicare reimbursed and the more profitable scenario. In this situation, nursing homes continue the care given in the hospital with emphasis on rehabilitation and recovery. The second type of situation in which nursing homes play a role is long term care. Generally, the patient is not expected to get better.[5] The success of long term care for the resident is judged by a slowing in the rate of decline in his or her condition. While important for both types of patients, a quality living situation is more crucial for patients who will be in the facility on a long term basis than for those in the more temporary post care situation. In any case, the staff of a nursing home or similar type long term care facility needs to have all of the skills necessary to care for both of these patient populations.

If it were up to them, the vast majority of patients with dementia would prefer to live in their own home where they feel they can retain autonomy and find comfort in the familiarity of their surroundings and life routines. There can be other advantages to living at home. For example, taking away the negatives of institutionalization such as being told when to eat, sleep and go to the bathroom may lessen the "negativity" and "behavior problems" of many patients. Family members of some patients with dementia feel more comfortable in making their relative's home as safe as possible even though they know that living at home might result in some risk of injury to their loved one, in order to postpone placing them in a nursing home.

Famous people with Alzheimer's disease have generously shared their experiences, demonstrating that even the most brilliant, talented and beautiful people are not immune from the terrible mental and physical consequences of the disease. Because of their stories, more attention has been focused on the need for research and, as we have seen, there are many clinical trials underway. However, the slow progress underscores the fact that there are inherent difficulties in finding a treatment for a disease in which the cause or group of causes is not completely understood. Even if a treatment is discovered, it will take many years before the effectiveness of the new medication and the extent of any adverse effects is truly known. Meanwhile, there will be an ever increasing number of people who will need help.

In my opinion, united support and advocacy has been lacking for people with dementia, particularly for those residing in nursing homes.

While it is true that more attention is being given to the rapidly increasing number of people with dementia, much of the concern has been driven by the large financial burden on the rest of the population rather than the actual plight of those battling the disease.

There has been growing attention and concern for family members and friends who are caring for people with dementia and realization of the tremendous physical and emotional toll it takes on them, although some of this concern occurs in the setting of fundraising or selling a service to the family.

Less attention has been focused on the patients themselves who are spending their last days, months or years in a nursing home facility. As their illness progresses, people with dementia lose their memories of the past and appear to have no real concept of themselves with regard to the future. However, they can certainly feel, experience and react to things in the immediate surroundings. Unfortunately for many, their life in the nursing home consists of the regimentation and indifference, and occasionally even abuse (subtle or not), by the nursing home staff.

Previous efforts to reform nursing homes have resulted in some minor changes and reforms in response to complaints but seem to have missed the "big picture." In terms of patients with dementia, nursing homes have not proven to be the best model of care. Care at home, or a situation as close to it as possible would provide better living conditions and fewer restrictions on patients.

In the meantime, there are, undoubtedly, (at least, I'd like to believe so) high quality nursing homes, assisted living residences and dementia care units with properly trained, skilled nurses and aides who perform their duties with patience and kindness. If you're loved one is not in one of these facilities, things can go very wrong.

Personal Stories

You may have noticed that nursing homes and other similar type of long term care facilities often have names that contain words like *village, palms, gardens,* and such. You would think that you would be checking your loved one into a luxury hotel. The marketing people and directors of these facilities give you a feeling of warmth and reassurance and use terms like *team, support, enrichment* and *home-like environment*. With their vast experience and big hearts they will take good care of your loved one. Well…maybe not. The narratives that follow tell a different story. (The identities of people and names of places have been changed.)

Sarah: *I was married to Lou for 40 years. He was a biology teacher and I was a lab technologist. One day, Lou said his right hand was numb. He sounded confused and he looked terrible. His doctor said to come right over. By the time we got to the office, Lou's numbness was gone. He was referred to a neurologist and had a lot of tests which all came back normal. The neurologist did not have a specific diagnosis. He told us it could have been an atypical migraine, a ministroke, or something else, but he really didn't know. He said the words I will never forget: "It will probably never happen again."*

Everything was good for about three months but then it did happen again---and again-- and again. Lou was also becoming forgetful. The episodes got longer and became more frequent over the next ten years and his memory got really bad. The doctor had shown me tiny spots in what he called "the deep white matter" portion of the brain on his first MRI exam. We watched these get larger and more numerous over the years as Lou got worse. No diagnosis was ever pinned down by the specialists. They told me it could have been something called multi-infarct dementia from repeated ministrokes. Later on they started calling it Alzheimer's. Nowadays, I hear they have tests for Alzheimer's disease. I once asked which one it was but by that time, they said it really didn't matter. Except for trying to keep his blood pressure and cholesterol down, they really didn't have any treatment. They did put him on Aricept but I didn't see any improvement.

Lou gradually lost the ability to do more and more things. He forgot the names of things and of people, even people in our own family. Sometimes he wanted to leave the house - even in the middle of the night - and became agitated when I stopped him. His diet became limited because he choked on liquids and some food. He needed help to do a lot of things like bathing or dressing.

Over the years, I cut down my hours and worked part-time so I could be home more. I hired people from home care agencies to help out when I had to leave the house. Speaking to the heads of these agencies and looking at their promotional material, you would think that their caretakers were all gentle, caring people who smiled a lot. From my experience, this was the farthest from the truth in a lot of cases. Most of the so-called caretakers that I encountered lacked any adequate training or knowledge about caring for elderly patients with dementia and some were sorely lacking in empathy. Lou could feel this and would try to leave or throw them out so I ended up having to leave work on many days. Lou's legs became weaker but he couldn't get the hang of using his walker properly. I got a wheelchair but he still tried to walk on his own and started falling.

Later on, I resigned my job to become Lou's caretaker and advocate. He couldn't bathe or dress himself and eventually couldn't feed himself. He was unintelligible and totally confused. I loved him so much and wanted to be with him. I could see the pain and confusion in his eyes sometimes. My heart was breaking. I didn't want to leave his well being and happiness in the hands of strangers. Lou started getting weaker physically and started falling a lot. Between the wandering and the falling, I was afraid to fall asleep. The 24/7 kind of care and help that Lou needed was prohibitively expensive. Lou's neurologist asked me if I "had a plan." His primary doctor, Dr. Lester Kare, said that Lou's admission to a nursing home was "inevitable." I knew there was no good ending for our situation. It was like we were standing in an open field with cannons firing at us and we had no place to run or to hide.

One day, Lou developed a bad urinary tract infection and had to go into the hospital. Lou's doctor said in effect that this was a good thing because the hospital admission would make it easier for him to gain admission into a skilled nursing facility. He could also have physical therapy there and help with his swallowing problems. I felt that there was only one path left to take and that I was being forced onto it.

While being treated for his urinary tract infection in the hospital Lou was evaluated by a neurologist and referred for EEG, CT scan and MRI. Since he had these exams multiple times in the past, I thought this was unnecessary but I gave consent anyway since they were, for the most part, harmless and, frankly, because they would break up his day in the hospital. Predictably, nothing new was seen on these exams; nothing that would enlighten the doctors or change Lou's treatment. Lou spent a good deal of time trying to get out of the hospital bed (where he didn't really need to be) and was very successful at maneuvering himself out between the side rails or off the bottom of the bed no matter how they adjusted it to keep him from doing so. I hired a health aide from a private agency to be with Lou from 8PM to 8AM and I covered from 8AM to 8PM. I had hoped to find a spot in an assisted living facility but Dr. Kare said he could not O.K. that, and I had to agree that from what I had seen there, these facilities could not meet Lou's needs.

I had to start visiting nursing home facilities in the area in earnest and I alerted the nurses to my short trips out of the hospital. Upon my return, I inevitably found Lou tied into a special chair by the nursing station, looking trapped and miserable. I was grateful that the male night aide could wheel Lou around and I tried to do the same during the day when I could get lifting help from the nurses or aides. I never laid eyes on the neurologist assigned to Lou in the hospital, and I always wondered whether she ever laid eyes on him, yet she wrote her notes in the medical chart and was appropriately reimbursed by Medicare according to Medicare payment notices I later received.

There are websites which rate the nursing homes, but for me the choices were really limited to say the least. Some of them had waiting periods of up to two years and others were too far away. I eventually chose a skilled nursing facility called Green Gardens. It had average reviews but I chose it mainly because Lou's physician worked there and would be familiar with his medical history and needs. It was also close to my home so that I could visit daily. One day during Lous's hospital stay, the saleslady from Green Gardens told me a bed was available for Lou and he could be transferred that very day. I had stressed Lou's propensity to wander and had requested a room near the nursing station and she assured me that this would be the case. It took a long time for the paperwork to be ready and it was dark when the ambulette came for him.

When Lou arrived, he was admitted to a room down a long corridor near an exit door, the farthest room from the nursing station! The staff told me that they had to give the room originally assigned to Lou to a woman who was being transferred from assisted living. I wanted to stay with Lou the first night but I was told that this was impossible because it was a two bed male room. After some intense negotiation they agreed to let the male aide, who stayed with Lou at night when he had been in the hospital, be with him that first night.

When I arrived early next morning, I noticed a sign on the door of the nursing home stating that there was a respiratory infection affecting patients on the floor to which Lou had just been admitted. It had been reported to the Health Department the sign said. I hadn't seen that sign in the dark the prior evening although I don't believe I could have rerouted the ambulette back to the hospital or to my home even if I had. Why would they admit a healthy but frail elderly person into a facility where bronchitis or pneumonia was going around? I called the medical director of the facility, Dr. Stuart Pidd, but Dr. Pidd seemed to have no idea what I was talking about. He said he had no knowledge of a respiratory infection problem in his facility.

The next morning, after a short meeting with the risk manager at my request, Lou was moved to a room that was not in sight of the nursing station but somewhat closer to it. His roommate was bed bound, although they did sit him up to eat, and he was very quiet. As to the respiratory infection, I was told that the problem was actually in the other wing of Lou's floor and it seemed to be getting under control. There was a bottle of sanitizer and masks at a station set up outside Lou's wing. Some of the personnel wore masks while other didn't. I suggested that there should be sanitizer outside or in each patient's room but the staff made it clear that they felt they had the situation under control.

A few days later, a care conference was scheduled at which time plans for physical therapy and feeding strategies to avoid aspiration were discussed. At the time, I also asked how I could arrange for a dentist for a general checkup for Lou. All seemed O.K. and everyone there "talked a good line," although none of Lou's nurses or aides were at the conference.

I missed Lou and wanted to spend as much time as I could with him. My daily visits confirmed prior fears and I got to see nursing home life from the inside. The patients all wore diapers and were taken to the bathroom at set times. Sometimes Lou seemed agitated and I sensed that he needed to use the bathroom but it was very hard to find someone to help me as I could not support him myself. It was hard to tell if they were understaffed and stretched too thin or were otherwise keeping a low profile.

Meanwhile, Lou's nurse, Nurse N. Charj, complained to me that he kept trying to get out of his wheelchair and because of this she could not give out her medications. I guess she couldn't find an aide to help her out either! I started to dread the look she flashed me when I arrived at the facility. Lou became one of a group "troublesome" patients who were positioned near the nursing station to be watched. I brought a CD player and some of our favorite CD's and wheeled Lou to the activities/dining area to give him some entertainment and mental stimulation.

Ultimately, Lou was put on Risperdal*, which they told me is an antipsychotic drug which was also being used in patients with dementia. Dr. Kare told me that he was going to give Lou the smallest possible dose. However, Lou became oversedated. Even the dentist told me that he managed to examine Lou's teeth but that Lou was too sleepy to have them cleaned. Lou was also too tired to do the physical therapy. Dr. Kare took him off the drug.

Organized activities at Green Gardens were few and far between. I wheeled Lou to one of these where a man sang and played the piano and the "inmates" were given ice cream and punch. One woman got up and did her version of dancing to the music and seemed to be having a good time. Everyone else just stared blankly and some fell asleep.

Lou had the ice cream and clapped weakly to a familiar song but soon lost interest. In my opinion the room they used for the event was too big to have any real effect at engaging the audience. An activities chart hanging on the wall of Lou's room and also by the nurses' station listed a schedule of room visits by the activities staff. These never occurred, at least for Lou's room. When I asked why, they said it was because of the infection going around!

Meanwhile, Nurse Charj was getting more flustered, if not annoyed, with Lou's behavior. Instead of trying to help him build up his muscle strength, the goal of physical therapy became to devise a chair that Lou could not get out of without the staff having to tie him in with restraints. They managed to do this by making the seat very deep with a steep incline for his legs, making it just about impossible for him to stand by himself.

The last straw for the staff came when Lou was found wheeling himself out of another patient's room with a slight cut on his hand. Nurse Charj notified Dr. Kare, who put Lou back on the Risperdal and they called a psychiatry consult. The nurses were pushing hard for medication. They told me that Lou really belonged in a locked dementia ward which they had done away with when they downsized some time ago. If he remained "exit seeking" he would have to be moved to a dementia unit at another facility and that could be anywhere, possibly too far to visit. I could feel my blood boil as she was threatening me with a concerned smile on her face.

---

* Risperdal is an antipsychotic medication. It has been used in patients with dementia and behavioral problems but that is an off-label use. There is an increased risk of death when used in elderly patients.

As a big favor to the nursing staff, the psychiatry evaluation was to occur the same week. I was told that it ordinarily takes much longer to schedule the psychiatry visit. The first part of the psychiatry evaluation consisted of a social worker examining Lou and getting a medical history intake from me. I had to sign a form agreeing to treatment before I knew what the treatment was and before the psychiatrist even laid eyes on Lou. They would not tell me when the psychiatrist was coming or schedule a meeting for me with him.

Since I was at Green Gardens every day, it was easy to ambush the shrink, Dr. Drew Bizzi. He admitted that Lou was too sleepy to evaluate but instead of eliminating the drug or trying something else, he continued him on the Risperdal and ordered an additional drug called Ativan which the nurses could use if Lou became agitated at night. Upon my protests, Dr.Bizzi said he was going to taper Lou's medication, but even then, I felt that he was just placating me. I began to wonder about Dr. Bizzi's credentials and looked him up on an internet site. It seems that Dr. Bizzi was a foreign medical school graduate which is not a bad thing in itself. Although he had a psychiatry residency in this country, there was no record that he was certified by any specialty board recognized in the United States.

I started to try to think of other options. Things were going badly at Green Gardens. I spoke to the relatives of other patients who shared similar stories but admitted that this was their only option. Still, I was filled with anxiety. Should I try to find another place? It had only been about three weeks: shouldn't I give this more time? Maybe it took time for them to develop an optimal program for Lou and to find a better medication, if that what was needed. I seemed to be watching an impending disaster but was powerless to stop it. I tried to speak to Dr. Kare but he seemed to be trying to avoid me. Did he feel guilty about Lou's course in the so called skilled nursing facility? Was he too busy with his private practice? Both? I could kind of understand if he was more interested in his private practice patients but his avoidance and lack of compassion were inexcusable. There was no guidance and no encouragement.

One day, Nurse Charj called me early in the morning to say that Lou had a "little cough" and that Dr. Kare had ordered an antibiotic. When I arrived at Green Gardens, Lou seemed even more tired than usual and had a bad cough and a low grade fever. I left late in the afternoon and asked the nurse to keep her eye on him. At 10 P.M. I called, as I had done every night since he was admitted, and the night nurse told me she had just looked in on him. He had "eaten his whole dinner" and she didn't hear him cough. I didn't see how this was possible but I assumed he was O.K. At 11P.M. she called me to say she had looked in on him again. This time he was coughing and had a fever. She heard something in his lungs and his oxygen saturation was low. She called Dr. Kare who told her to have him admitted back to the hospital. I rushed to the emergency room where an IV was already put in his arm and he had a catheter in his bladder. They also gave him oxygen.

When I got to the emergency room and first saw Lou, he looked like he was about to die. I knew then that the night nurse hadn't really checked on him at 10P.M. and if I hadn't called he might have been dead by morning. They did a chest x-ray which showed he had pneumonia and they took blood. Finally, they managed to get Lou stabilized. While they were reviewing his chart, the ER staff told me, that contrary to what I had been told, Lou had been on a **high** dose of the antipsychotic drug. The pneumonia could have been from an infection or might have been related to aspiration.

By now it was 2AM and Dr. Kare was not answering the emergency room's calls and pages to get the admission orders. After a couple more hours they finally got hold of Dr. Kare and Lou was formally admitted to the hospital. He languished there for about a week. He had a horrible rattling cough and his chest x-rays did not improve. He got weaker and weaker. The speech therapists advised that it would not be safe for him to have any food or meds by mouth due to danger of aspiration, not that he looked in any shape to take food. The respiratory therapist came daily but it soon became clear to the nursing staff that there was really nothing they could do for him.

Lou's deep rattling cough and his general condition were getting worse. He was agitated and constantly pulled off his nasal oxygen. Later he ripped out his intravenous line and bladder catheter. I was desperate. He was already in a hospital. What was there left to do? Where was Dr. Kare for guidance? After a while I did hear from Dr. Kare, who told me that the only way Lou could get nutrition would be to put in a feeding tube. What did I want to do?

From my readings online and from speaking to other doctors I knew and trusted, I learned that the feeding tube could, in itself, lead to greater problems and it was becoming clear to me that it was not the right path to take. At the very least, I was certain that Lou would pull it out, leading not only to possible medical complications but also to more sedation, more physical restraints put *on him and more misery! Any which way, he would never have the life I knew would be acceptable to him.*

Lacking any good options, and without much hope, I began to wonder if hospice or palliative care in the hospital would be appropriate and asked Dr. Ernest Hart who was also attending to Lou. He said "yes" and a consult was arranged. Lou was deemed to be at the end of life and was admitted to the hospice. There, he spent the last week of his life in peace and dignity without any visible signs of pain or agitation and where I could be with him 24/7. I wondered why Lou's doctor, or anybody else for that matter, hadn't discussed this option with me earlier.

Sarah's story is not unique and it tells you just about everything that is wrong with institutionalized care of people with dementia. She and Lou experienced many of the known, longstanding problems faced by dementia patients and their families when long term care in a nursing facility becomes necessary. When the needs of the individual, who is in the later stages of Alzheimer's disease or other type dementia, exceeds the capabilities of even the most dedicated family, options become extremely limited or even nonexistent.[6] Sarah relates in her narrative that Lou's physician informed her that nursing home placement was "inevitable".

All too often, however, nursing homes serve only as "warehouses" for the patients where the term "quality of life" is not even applicable. The patients must eat, sleep and wake when they are told which is really antithetical to people with dementia. They are all put in diapers which are changed at certain times for the convenience of the staff. In Lou's case, the nursing home director was undoubtedly collecting a salary but totally detached from what was going on. The medical care decisions by nursing home physicians are often driven by the nurses. Lou's nurses did not have the time and they lacked the training to deal with his confused wandering. It was easier to subdue him with medications. Although tying the patient to the bed with restraints is "frowned upon," the staff gets around this by doping up the patients with medications which are known as "chemical restraints." Interestingly, Lou's condition was ultimately called Alzheimer's disease. Although his diagnosis was never pinned down specifically by any medical tests, it didn't matter for his fate would have been the same.

Except for treating any obvious reversible causes of dementia, the same holds true today.

"All the king's physicians and all the king's medicines" cannot put the dementia patient back together again. All the support groups in the world cannot repair the utter feeling of desperation and helplessness experienced by the families of patients with late stage Alzheimer's disease and other advanced dementias. Nursing home placement often becomes the best of several really bad options, or may be the only option. The following excerpts reflect the anxiety and sadness experienced by many caregivers who have had to make this decision.

Leslie: *I had to put Mom in a nursing home, I have three kids and I work and she couldn't be left alone. I moved her in with us for a while but it just wasn't working either. I visited her a couple of time of week in the home, as often as I could. I feel bad for the residents who don't have relatives to visit them. They seem to be left on their own, wheeling themselves down the hallway. Occasionally I've helped someone who got stuck trying to get through a doorway in their wheelchair. Except for simple instructions like "O.K., stand up now, lie down, get back in your chair"- things like that- when the CNAs are feeding or dressing them, they don't really seem to try to talk to the patients. Some of them are just placed in their wheelchairs by the nurses' station. I guess they get to see some activity but it's sad that that's all there is to their life. Once, I surprised Mom and had breakfast with her in the dining room. The CNAs were talking to each other but didn't really talk to the people they were feeding. There was music playing softly but it sounded like rap. It certainly wasn't anything my mom, and I'm pretty sure the other people, could relate to.*

Joe: *I wheeled Dad to a party they were having in the dining room. The entertainer played old songs on the piano and then on the guitar. Cake and ice cream were served. Two or three people clapped along to the music. One lady sat up very straight in her chair and had a slight smile and one elderly man looked like he was paying attention. The others just stared expressionless and some were sleeping. None of the staff was helping them or encouraging them to eat or to participate. Mostly they all looked zonked.*

Joanne: *There's a strange sameness to this place every time I visit Mom. There's a lady who wheels herself, up and down the corridor; up and down, up and down. She is bent all the way over. It always make me think: Is something wrong with her spine or does she prefer to view her world from that position? Another lady with some kind of combination chair and walker, smiles and points at you every time you walk by her. Another patient yells something like "yah, yah, yah" if you enter her space. I don't know if she is scared or just excited. Another woman with long gray hair wheels herself down the hallway, always asking anyone she sees where her room is. The staff ignore her but first time visitors try to help. Then there is one woman who can walk on her own who comes out of her room, then goes back in, then comes out, then goes back in. Well, you get it. It's all kind of depressing but I've been coming here for several months now and I guess I would be upset, if one day I didn't see one of them.*

Sherry: *I sometimes wonder about the training these people have. I started talking to one of the nurse's helpers. He told me he's a construction worker and he's supplementing his income. He rarely has a day off and he's been working double shifts – 15 hour shifts. That can't be good.*

One day when I came to visit my aunt, I found a blood stain on her pillow and I asked the nurse's aide - a different one- who was on duty that day to change the bedding. I stopped myself from asking the question I really wanted to ask: "How the hell did that get there?" Even so, instead of apologizing or telling me she'll try to find out what happened, the aide got very defensive and looked angry. I didn't want her taking it out on my aunt so I tried to make some friendly conversation with her. She calmed down and got a little friendlier when it turned out that we were from the same town in New Jersey. Still, that show of temper had me worried.

Bill: *They're constantly getting people's clothes mixed up. One day I found Dad wearing someone else's pajamas and some of his stuff was missing. They have a room where you look for the missing clothing but it's pretty disorganized. I hope that they're not mixing up the residents' medications!*

Deanna, CNA: *It's always so busy on the floor. I can barely keep up. One evening I accidentally handed some woman another patient's tranquilizer. I* didn't tell anyone but I was scared it might kill her and I kept checking on her the whole night. Luckily she was O.K.

Cindy, CNA: *One day one of the patients wouldn't return to her room and she kept saying, "I'm waiting." I asked her why she couldn't wait in her room and she answered that she was "waiting at the dock of the bay." I couldn't stop laughing. These people are so funny.*

Stan: *My mom's left leg was swollen. I am a paramedic and I was worried she might have a clot. Her nurse told me, "That's impossible, her leg is not red." I know that it IS possible. It was all I could do to get them to do a sonogram to look for a clot. I think they just didn't want to be bothered.*

Marilyn: *The nurse kept complaining to me about Dad. She said he was "doing obscene things." With a very serious look on her face she told me that he touched a physical therapist's leg and a nurse's breast. I felt like I was in the principal's office. What am I supposed to do? Doesn't she know that his judgment and just about everything else that used to be in his brain is gone?*

The comments of these family members and nursing home staff illustrate the lack of sufficient staffing in terms of number, and more importantly, in terms of experience and training. Understaffing is particularly problematic in this setting because of the considerable care and help that dementia patients require.

The nurses need to have time to deliver medical care and dispense medications without distractions so that no errors are made. Yet there still needs to be time so that the residents can have their individual personal needs met. Even if there are enough "bodies" on staff to feed, clothe, bathe and medicate the patients, it is just as important to have staff members who will talk to them, engage them in activities and generally treat them as worthy human beings. Staff members need to know in their heads and in their hearts that when patients with dementia wander or become agitated, angry or scared for no reason apparent to us, it is NOT THEIR FAULT. When they say or do something inappropriate, it is also NOT THEIR FAULT.

PROBLEMS WITH LONG TERM CARE

According to the Centers for Disease Control, more than two-thirds of people aged 65 can expect to require long term care services during the remainder of their lifetime.[7] Long term care means the ongoing help given at home or at an outside facility that a person needs to perform the usual daily activities of living. This could involve having help eating, getting dressed, bathing, or going from one room in a house to another. It might only require a home health aide for a part of the day. Long term care might also involve catheter care or administering medications which might be done at an assisted living facility or nursing home. People with Alzheimer's disease usually require an increasing level of care and help doing things as the disease progresses.

Later in the course of AD, so-called behavioral problems can develop which may be very frustrating to caregivers. The erratic and inappropriate behaviors may be dangerous to the patient and those around him or her. Care for the individual at home may not ultimately be feasible. Some people believe that caring for a parent or spouse at home may not be as important once patients reach the stage where they no longer recognize their caretaker and family members and do not even know that they are at home. This is debatable.

Nursing home placement is usually thought of as the last resort (pardon the pun, because a resort it is NOT). However, placement of a loved one may be initiated because the individual has medical problems requiring skilled nursing care. Family members also consider nursing home placement if the parent, spouse or other loved one has become dangerous or disruptive, or if the caregiver is having difficulty in managing the individual and there is no one else to help out. Theoretically, at least, the nursing home may offer a safer, controlled environment. Incontinence is still another reason why some dementia patients are placed in a long term care facility.

There are, however, disadvantages to nursing home placement that should not be overlooked. For example, there is a real risk that the level of care and the quality of life that the individual receives at a nursing home may be inferior to that received at home. At best, they will receive less individual attention. Some people decline rapidly after nursing home placement.

Angela: *I saw an article on the internet recently which made me sick. The son of a former resident at a nursing home in New York was suing the home after his 85 year old mother was subjected to a stripper act. There is a picture showing the stripper leaning over the man's mother and she's putting a dollar bill in his underwear. This is such an outrage, and I can't help but wonder if things like that can be going on in my mom's home when there are no visitors around.*

One can find many stories like the one above where aides have been caught humiliating, if not outright physically abusing, elderly nursing home patients with dementia.[8, 9, 10] This has led some family members to install hidden cameras, sometimes called "granny cams."[10,11] Some states, including Texas, Oklahoma, New Mexico, Maryland and Washington have laws allowing electronic monitoring in nursing homes. The data could be used as evidence in cases of litigation by families and the presence of monitoring equipment could serve as a deterrent to bad behavior by the staff. Of course, the secret monitoring of nursing home residents raises issues of legality and ethical concerns. Some have proposed notification of the facility of the use of cameras with the hope of deterring bad treatment of residents. Of course, emphasis should be placed on finding ways to prevent the abuse of patients before it begins.

Let's look at some of the problematic aspects of care associated with nursing homes and related long term care facilities.

The Care

A "skilled nursing facility" is where many patients end up in order to receive rehab or specialized care after a hospitalization.[12] Medicare will pay for skilled nursing facility care for a specified period, after a medically necessary three day or more stay in a hospital for a related illness or injury.[12,13] Approximately 90% of the more than 15,000 skilled nursing facilities in the United States are also certified as nursing homes where patients get long term type care. Sadly, some residents do not survive the care given in these facilities. This has been traced to several common problems.

One issue is that many people in the advanced stages of Alzheimer's disease continue to get medications that, at best, are no longer beneficial to them and, at worst, can further damage their already compromised health and well-being. Findings of a study from the University of Massachusetts Medical School indicates that nearly 54% of nursing home residents with advanced dementia in 460 facilities were prescribed one or more medications of questionable benefit between 2009 and 2010. Often they are drugs used to treat AD that are no longer helpful in the late stages of the disease but which subject the patients to unpleasant and potentially harmful side effects.[14,15]

Even more startling is the report that one third of patients in skilled nursing facilities have been hurt from a medication, an infection or other issue related to their treatment. Doctors working at the Office of the Inspector General of the United States Department of Health and Human Services reviewed the records of 653 Medicare patients from more than 600 facilities who were receiving treatment in nursing homes after discharge from an acute care hospital. The bad outcomes for these patients resulted in lasting harm in 22% and temporary harm in 11% of the cases. In 1.5% of cases, the poor care resulted in death.[12] The causes of patient injury included blood clots, excessive bleeding from blood thinning medications, fluid imbalance and kidney failure. There were cases of inadequate monitoring, substandard or delayed treatment and even instances total failure to provide necessary care. According to the physician reviewers, 59% of the errors were preventable.[12]

The Staff

Who is actually looking after the residents in long term care facilities and what is their level of training?

Many facilities will have a registered nurse (RN) for at least part of the day. They have completed a two year nursing program and have, at the least, an Associate's Degree in Nursing. RNs receive certification through an exam. A BSN is a registered nurse that has earned a bachelor's degree. The mean salary of an RN is $64,886 and that of a BSN is $73,091. Salaries vary with experience and location. Registered nurses with less than a year experience may earn $38,521 to $53,548 yearly while an RN with more experience may earn between $50,426 and $72,076 yearly.[16,17]

Licensed practical nurses (LPNs) complete a one year nursing program and also must pass an exam to be certified. They earn a mean salary of $42,557, again depending on how much experience they have and where they are working.[16] They may do many of the things that RN's do in terms of patient care and record keeping. A long term care facility may have an LPN instead of an RN.

The people who make up the majority of the staff at the long term facilities are the certified nursing assistants (CNAs). Each state board of nursing has different standards for nursing assistants or nurses' aides. In order to be registered or licensed by the state, an applicant must pass an exam administered by the state. Usually the states require a criminal background check and there may be requirements for continuing medical education. The CNAs perform the more routine nursing chores in order to help free up the RNs and LPNs. If you go on line or look at the newspapers, you can see ads for CNA training programs, most of them running anywhere from 2 weeks to 15 weeks, although I did find one advertised course for as short as one week and another for as long as 24 weeks.

In a nursing home situation, CNAs help feed, dress and bathe patients. They assist them with toileting and may help them with exercise as well. Some states allow CNAs to give out medications but they must go through an additional certified program. Often CNA programs are given in nursing homes and other elder care facilities. They may range from two to six or eight weeks and consist of formal instruction and practical experience. Trainees learn about patient hygiene, biohazards in the workplace, medical ethics and about some of the laws related to nursing care. They are also taught the signs of an impending medical problem. The nursing facility may pay for the training but, in turn, it gets a ready supply of CNAs. The pay for CNAs is low, at approximately $8 to $16 an hour, and the turnover rate is high.

The two largest employers of CNAs are nursing homes and elder care facilities, In fact 60% of all CNA jobs are in these types of facilities. The Bureau of Labor Statistics has projected a 20% increase in the number of CNA jobs in 2020. Currently, the training of CNAs can be measured in weeks or months, if not days, and experience in caring for the elderly patient with dementia is minimal or non-existent. Disturbing reports of cases of criminal neglect and abuse of elderly nursing home patients can be found in newspapers and on the internet.[8, 9, 10, 11]

Skilled nursing facilities, nursing homes and the like have medical doctors on their staff but you may not be able to find them there much of the time. The nursing home position may be an adjunct to their private practice and will always take second place.

We should insist that CNAs and others who are caring for those with dementia should have better training. Reimbursing them adequately would also go a long way in attracting more qualified people.

The System

Late stage dementia patients are frequently bounced around, back and forth in the system, between the hospital and the nursing home, often much to their detriment.[18, 19, 20] In one common scenario, a resident in a nursing home may develop pneumonia due to aspiration of food and saliva, a problem frequently seen in patients with dementia. They are transferred to the hospital where they frequently end up with feeding tubes. All this stresses them further and they are given medications to calm them down and restraints to prevent them from pulling out the feeding tubes. The feeding tubes themselves, whether in place or dislodged, can give rise to other problems.[20]

Some experts believe that many of these hospitalizations are avoidable and that treatment could often be given at the care facility. Financial considerations are also factored into the decision making. After a three day inpatient stay at a hospital, nursing home residents who have Medicaid coverage may qualify for Medicare Part A payment for the post acute care in the nursing home. The nursing home will be reimbursed at three to four or more times higher than the daily rate paid by Medicaid![18, 19, 20]

The Cost

Long term care can be very expensive.[21,22] Various factors, including location of the long term care facility may influence pricing. The numbers quoted here come from the results released in 2013 of a biennial long term care cost study by the John Hancock Life Insurance Company. The study, carried out by LifePlans Inc., included 16,000 long term care providers in key cities across the country. Nursing homes, assisted living facilities and home health care agencies were surveyed. According to the results of this study, the average annual cost of care was $94,170 for a private room in a nursing home, $82,855 for a semi-private room in a nursing home, $41,124 for an assisted living facility and $18,460 for adult day care. For those receiving home health care, the average annual cost was about $29,640.[21] Not surprisingly, the study also showed that costs of care had increased in all categories.

What about long term care policies? They may cover expenses related to care at home or in a facility. There is a lot of variability in the cost of these policies, depending upon such factors as your age, where you live and your choice of benefits. Premiums quoted by the National Association of Insurance Commissioners include an average yearly cost of $888 for a 50 year old, $1,850 for a 65 year old and $5,880 for a 75 year old.[22] If you stop paying the premium, the policies will end in most cases and you lose all the benefits. Policies that do preserve some of the benefits will usually be higher priced and require that you will have had the policy for a specified number of years.

Medicare does not pay for long term care in the setting required by patients with Alzheimer's disease or other dementias who require housing and basic care. It will help pay for up to 100 days in a skilled nursing facility after a hospitalization meeting certain requirements. It will also reimburse for 100 days of skilled nursing services at home after a hospitalization but this does not include help with normal activities like eating or dressing.[13]

The 2013 World Alzheimer's Report urged leaders worldwide to make dementia a priority and devise strategies for long term care as the expected increase in patients with dementia would require greater governmental support for the current care given by family and friends in a dedicated but informal way. Currently, four-fifths of the money spent by Medicare is on individuals with Alzheimer's disease or other types of dementia. On average, Medicare spends three times as much per person on those individuals with dementia including Alzheimer's disease, as compared to others.[23] Medicaid spends approximately nineteen times as much per person on those seniors with Alzheimer's disease and other dementias as compared to others.[23]

As part of a new trend, in an attempt to limit the increasing Medicaid cost of long-term care some states began instituting mandatory programs that puts public money into privately managed long-term care plans. The new system was supposed to provide alternatives to nursing home placement and allow people to remain in their own homes longer.[24,25]

The problem with this plan, however, is that the pressures of profit motive and cost containment inevitably seem to win out and have led to stories of denial of care as people's needs became more costly. In many cases, individuals lost care that they would have been entitled to under the old system. For example, formerly people who would have gone into a nursing home, with Medicaid bills from state and federal funds were now able to receive daily help for a set number of hours at home with the option of going to a nursing home later, when necessary. Medicaid paid a set monthly amount to an insurance company which would cover and coordinate the person's care. The patients preferred to be at home and the government and insurance company profited, too, because the care at home would be much less than the monthly amount that Medicaid paid to the plan, less than a nursing home would have been. However, as their needs increased, there were reports that people lost their home care and were also denied nursing home placement. The reason for this was that the disability requirement to be admitted into a nursing home or to get the equivalent care at home was raised sharply.[25]

What's behind all this? Approximately 4.2 million people receive long-term services which are paid for by Medicaid to the tune of approximately $136 billion dollars. This represents one-third of all Medicaid spending but only 6% of Medicaid beneficiaries. Under the prior way of doing business, providers bill Medicaid which leaves the system vulnerable to not only more expensive care but also unnecessary care. This was to the advantage of the nursing homes. Spending would be more controlled under the new system of managed care. Plans pay networks of providers of care from a fixed amount per enrollee. The expense of the people who need more care would be balanced out by those people who don't need as much care.

However, the newer system has led to other kinds of abuse. While the model of paying a set monthly fee to cover the cost of each individual helps state officials to predict and limit Medicaid costs, critics are concerned that insurers are incentivized to increase profits by skimping on benefits. There were found to be plans that preferentially picked relatively healthy seniors who should not have been eligible while turning away the most disabled elderly, among whom were bed bound patients with dementia.[25]

Quality of Life: Keeping the Patient "In Line"

Patients with Alzheimer's dementia have increasing loss of memory, inability to communicate with language, as well as loss of judgment and other cognitive abilities. They need more and more help from others to perform the routines and tasks of daily living. Further on in the course of Alzheimer's, there are often additional symptoms including irritability, agitation, and loss of inhibition. These disturbing manifestations of the disease often cause the patient to be removed or isolated from social situations. The abnormal behavior is well known to negatively impact the lives of family members and can seriously disrupt the whole household. Sadly, it is often a major reason why those with dementia are placed in nursing homes or other similar type institutional facilities.

A nursing home or other type of long term facility is generally not the first choice of the family for their loved one, but they are forced into this decision by lack of better options. By the time the patient is in need of such services, he or she usually has no say in the matter. It is reprehensible that many dementia patients are placed in facilities that make promises that they cannot keep. There is a reason why many people say that, if faced with the decision, they would choose suicide over placement in a nursing home. Life for a nursing home patient with dementia can be especially grim. At best, it is depersonalizing and nonstimulating with all the negative aspects of being" institutional". It can also be very frightening and at times demeaning. Because many nursing homes are understaffed, it is easier for personnel to treat their residents in batches or in bulk. They eat at the same time and have their toileting needs addressed at the nursing home's scheduled times. This is because the nursing home would not have enough staff to take care of an individual's impromptu needs, much less wishes. However, often patients with dementia do not do well with this type of regimen.

What looks like agitation may be induced by pain, an unrecognized infection, the feeling of hunger or the need to go to the bathroom but the patient has lost the ability to express their needs appropriately. The harried staff may only see these signs as an impediment to their daily work schedule. "Good patients" are those who are docile and submit to the regimen. Patients who do not comply are looked upon as 'bad" or troublesome. It takes a caretaker with training, experience and empathy to truly understand that the patient is really unable to act appropriately or reasonably because they have dementia, not because they are out to give the staff a hard time.

The Use of Chemical Restraints

The use of physical restraints on patients is discouraged and the practice is noted negatively by those agencies that inspect nursing homes. Tying a patient to their bed or wheelchair certainly doesn't look good for relatives and guests. Sadly for the patient, however, there are ways to get around this prohibition. As we saw in Sarah's story, the overriding aim of the physical therapy sessions was to improvise a wheelchair in which her husband was pretty much trapped, without having the appearance of him being tied into it with belts or straps.

An even more despicable tactic is the indiscriminate use of "chemical restraints". The "difficult" patients are given drugs to make them more manageable. To date, the U.S. Food and Drug Administration (FDA) has not approved any medication for the purpose of treating agitation associated with dementia. That is not to say that a host of pharmaceuticals has not been tried on these patients. These include the drug categories of cholinesterase inhibitors, antidepressants, anticonvulsants and antipsychotics, among others.[26,27]

This unfortunate practice was addressed in the 1987 Nursing Home Reform Act which prohibited the use of chemical restraints on patients.[28] Later on, in 2005, the US Food and Drug Administration warned that use of the newer generation atypical antipsychotic drugs in dementia patients increased their risk of death. Clinical trials showed a 1.6 to 1.7 fold higher death rate in those who had been prescribed this type of treatment compared to those who had been given a placebo. The FDA issued an advisory and "black box" warning* related to the use of these atypical antipsychotic drugs in the elderly with dementia.[27-30]

There have also been legal challenges to the use of antipsychotic medications for patients with dementia. The pharmaceutical company, Johnson and Johnson has received considerable, and well deserved, criticism and has been embroiled in numerous legal actions over the improper marketing of its drug Risperdal (generic name, Risperidone). [31-35]

Charges regarding improprieties involved with Risperdal, led to a $158 million settlement in Texas.[31] In 2012, a $1.1 billion judgment was imposed on Johnson and Johnson for their improper marketing of Risperdal, and for not revealing the risks of this medication. The company had been sued by Arkansas for violating the state's Medicaid fraud law and its deceptive trade practices act. Johnson and Johnson stood by the efficacy and safety of their drug, Risperdal, which is used in the treatment of schizophrenia and bipolar disorder. In fact, Johnson & Johnson had been defending its marketing practices of Risperdal for years. According to prosecutors, however, the company did not disclose the dangerous side effects of the drug such as a higher risk of diabetes and weight gain as well as an increased risk of stroke in older patients. Johnson & Johnson appealed the judgment and it was overturned by the Arkansas Supreme Court in 2014.[35] The reason for this outcome is that the court reached the conclusion that the law does not apply to drug makers like Janssen, which is Johnson and Johnson's pharmaceutical unit, because it is not a health care facility. Pharmaceutical companies in other similar circumstances have been able to win at the state level because many state laws were instituted to protect the public against the improper actions of health care providers, but not necessarily from the drug companies.

Other legal actions were also reversed on appeal. One such example was that of the Louisiana Supreme Court which overturned a $258 million judgment claiming that not enough evidence was produced by the state prosecutors that the company's marketing of Risperdal was improper.[31]

---

* Black box warning - the strictest warning put on the labeling of prescription drug products by the Food and Drug Administration (FDA) when there is reasonable evidence of an association of a serious hazard with the drug.

Legal battles went on and Johnson and Johnson was sued by eleven state attorneys general alleging that the company mislead the public about the safety and efficacy of Risperdal to boost its sales.

Finally, in late 2013 Johnson & Johnson and its subsidiaries were fined over $2.2 billion to settle federal claims that it had improperly marketed its drug, Risperdal, for use in the elderly, in children and in individuals with developmental disabilities.[34] The claim was that the pharmaceutical company minimized the potential for serious risks when the drug was used in the elderly and that the medication had not been approved for the elderly population. Johnson and Jonhson's aggressive marketing practices were also alleged to include kickbacks to doctors and to Omnicare, the largest long term care pharmacy provider. Johnson and Johnson did not admit to wrongdoing in response to these claims but it did agree to plead guilty to a criminal misdemeanor acknowledging that it marketed Risperdal improperly to older adults for unapproved uses. The settlement was the third largest pharmaceutical settlement in United States history.

As far as the medical risks, a recent study, published in the Annals of Internal Medicine, found that acute kidney injury is still another potential hazard to those elderly patients given antipsychotic medications in an attempt to treat behavioral problems, such as agitation and aggression, associated with their dementia.[36,37,38] Besides risperidone (Risperdal), these drugs include quetiapine (Seroquel), and olanzapine (Zyprexa). These medications are FDA approved for treatment of mental disorders such as schizophrenia and bipolar disorder.[36] While deleterious effects of the off- label use of these drugs in elderly patients with dementia have been previously reported, this newer study revealed that the elderly taking one of the three drugs were one and a half times more likely to suffer acute kidney injury as their peers who were not taking the medications. The risk of low blood pressure and an inability to empty the bladder (otherwise known as acute urinary retention) was twice as high as in those patients who had not been prescribed the medications. In fact, the risk of death from any cause was more than twice as much for the elderly given these drugs.[36] Apparently, these medications may be effective for adults with true psychosis. However, it is generally acknowledged that they should not be used indiscriminately, as they are now in many institutions.

Patients with dementia who cannot communicate their physical and emotional needs effectively may act inappropriately. Sadly for them, it is easier for the staff of long term facilities to medicate them than to take the time to find out if they are hungry, cold, afraid or have some simple problem that can be remedied.

Unfortunately, federal efforts to reduce the practice of using atypical antipsychotic medications on elderly patients with dementia were stalled. This was attributed in part to budget cuts that prohibited adequate investigation of the problem but also to the failure of any voluntary effort to meaningfully address the problem. In May 2012, The Centers for Medicare & Medicaid Servives (CMS) created the Partnership to Improve Dementia Care.[39,40] There was to be involvement by leaders at nursing homes, doctors, pharmacists, and drug manufacturers and the goal was a 15% reduction in prescribing of these antipsychotics to nursing home residents by the end of the year. There was an exclusion in the case of some diagnoses including schizophrenia and Huntington disease where these medications may be indicated. However, there was only a 9% reduction in nursing home residents receiving these drugs between the fourth quarter of 2011 and the first quarter of 2013.[39]

High rates of prescribing of atypical antipsychotic medications to the vulnerable population of the elderly continues. In fact, data from the Centers for Medicare & Medicaid Services (CMS) showed that more than one in five nursing home residents were given these antipsychotic medications in the first quarter of 2013 even though the residents did not have a diagnosis that would justify their use.[39] In addition, according to a report by of Office of the Inspector General of the Dept of Health and Human Services, 99% of all nursing homes tracked failed to meet federal requirements for putting into place periodic evaluations and care plans for those patients taking the antipsychotics. These steps were intended to check that the patients have a diagnosis that warrants taking antipsychotic medication and that behavioral interventions were tried in order to be able to lower the dose or perhaps even eliminate the need for the drugs. Review of nursing home records of 375 residents by the OIG revealed that only two of them met all of the federal requirements![39] CMS has added the use of antipsychotic drugs to the quality metrics in the agency's Nursing Home Compare database in an effort to increase the accountability of nursing homes but deficiencies in enforcement have been noted. An example given is that harm citations were not issued when inappropriate use of the antipsychotics was encountered. Some individual states have brought the levels of inappropriate usage down but much more has to be done in this regard.

The use of these antipsychotics is often driven by the nursing home personnel who are inadequately trained in caring for the elderly population with dementia. It's a given that people with dementia can be much more difficult to care for than those with other medical conditions. The nursing home staff in these institutions may not say it directly but obviously display the need to have their patients more docile in order to get their work done. Actually, however, providing the best care for their patients should be the main component of the work they get done. They will also tell you that the drugs will help to prevent accidents in these patients. However, patients can be more susceptible to falls and other types of injury if they are overmedicated. Most importantly, the risk of death from unwarranted medication is unacceptable.

Recently, other medications have been tried for the treatment of the agitation often seen in Alzheimer's patients. In a randomized trial involving 186 patients with dementia and agitation, subjects were given either a drug called citalopram and psychosocial intervention or a placebo and psychosocial intervention for a period of nine weeks. Using established assessment scales, those who had received the citalopram showed less agitation than those who got the placebo.[41] However, two years after the start of the trial, there were concerns about adverse cardiac effects and a question of slight cognitive decline in patients taking the medication and an advisory was issued by the FDA.[41] A drug called dextromethorphan-quinidine is one of several other pharmacologic agents being studied as potential alternatives to the atypical antipsychotic medications.[42]

As appears to be the case for many studies related to Alzheimer's disease and dementia, results bring up more questions and uncertainty. There is a need for more studies of large populations of elderly AD patients with agitation, carried out over an extended period of time. Finding effective medications may not be a "one size fits all" situation. There could be evaluation studies with careful controls to find out which dose regimen is optimal and what characteristics in the patients are more likely to result in a successful outcome. Of course, patients should be carefully monitored for side effects. To date, however, there is no proven effective and safe treatment for the symptom of agitation in Alzheimer's disease.

What dementia patients require more than medication, and what most nursing home type facilities do not provide, is a more individually based type of care. These patients need to be protected and monitored but whenever possible they should be encouraged to be more active and engaged with their surroundings, as their overall condition permits. Furthermore, giving residents more autonomy would probably cut down on some "negative" behavior. For example, since the sleep-wake cycle is often abnormal in patients with later stages of dementia, they could be allowed to go to sleep and wake when they prefer. However, most institutions are not set up for this type of more humane approach

A Five Star Rated Nursing Home by Medicare May Really be a "Hell Hole"

With all the inherent faults of so many nursing homes, how do you know you're entrusting your loved one to a good one? You cannot base your opinion on a cozy looking setting or on the "spiel" given to you by a glib director of the facility. There are valid reasons to be wary. Remember that according to the report issued in 2014 by the Office of the Inspector General of the Department of Health and Human Services, 22% of Medicare patients who stayed in a nursing facility for 35 days or less were harmed by the medical care (for lack of a better word) they received. Aside from that group, another 11% sustained temporary injury. In 2011, Medicare spent approximately $2.8 billion on hospital treatment because of the harm inflicted on patients in nursing homes.[12,43]

Nursing homes have been rated using a scale of one to five stars on a Federal website called Nursing Home Compare.[45] Unfortunately, ratings and evaluations of these facilities may be quite misleading.[44,45,46] It turns out that Medicare ratings, upon which many rely, are based largely on data submitted by the nursing homes being evaluated and not confirmed by the government.[44] This is unbelievable!

Only the results of annual health inspections are based on evaluations of independent reviewers. Staff levels and quality statistics are the other measurements used to determine the star rating and, incredibly, these are reported by the nursing homes themselves. In addition, a history of fines, consumer complaints filed with state agencies and enforcement actions by state authorities are not used to determine the ratings. This is important because the ratings are often used to decide where to refer a patient after discharge from the hospital.

Meanwhile, the number of nursing homes with above average ratings was increasing. For example, 37% of these institutions received four or five star ratings in 2009. In 2013, nearly 50% of the homes got these high ratings.[44] This does not mean, however, that quality of care truly improved to that extent. Nearly two-thirds of fifty nursing homes which were on a federal watch list because of poor quality received four or five star ratings for the self reported categories of staff levels and quality statistics. The rating system was established in 2009 in the hope of greater transparency in the evaluation of nursing homes and, not surprisingly, at first the nursing home industry lobbied against the system.

Some patient advocate groups find the rating system misleading and only helpful in weeding out the "worst of the worst." The nursing homes know how to game the system. For example, to boost their scores in staff levels, some homes have added workers in the period leading up to an annual inspection. The Affordable Care Act of 2010 requires that staffing levels reported by the nursing homes be confirmed by Medicare using payroll information.[44]

In the other self reported category of quality measures, plans have been formulated to institute spot audits of the facilities. There is also a plan to add other criteria, such as the number of residents who are on antipsychotic medications. Early in 2015, it was reported that almost one-third of nursing homes in the United States had their ratings lowered. In my opinion, these measures are long overdue!

Summing it Up

Many of the elderly with Alzheimer's disease and other types of dementia end up being placed in nursing homes and other long term care type facilities by their desperate family members. There are deficiencies in the quality of care which almost seem built into the system and these facilities are very expensive. Elderly people with dementia are a very vulnerable population and may be at the mercy of inexperienced and occasionally abusive workers as well as ineffective and dangerous drugs.

VI FINAL THOUGHTS
Changing the Status Quo

So where **do** we stand in the fight against Alzheimer's? In the time since Alzheimer's disease was named, more than 100 years ago, scientists have learned more about the specific pathology of the disease and about the working of the brain in general. Imaging techniques have been developed to depict the brain in exquisite detail. However, the information gained from modern research has, to date, been insufficient to bring about any meaningful improvement for all the people with Alzheimer's who are waiting for help. Scientists have not been able to define, and cannot agree on, the exact pathological processes in the brain that bring about the symptoms. The focus of much of the research has been on the beta amyloid plaques and tau proteins which have been so strongly linked to the disease. However, the science is fraught with controversy and many theories have been espoused. Finding an effective treatment has, so far, also been elusive. While an association between the development of Alzheimer's and the presence of certain abnormal genes is known, many other predisposing factors have been postulated but are not yet proven.

Because of the numerous disappointing attempts to develop a treatment for Alzheimer's disease, more recently many researchers have been turning their attention toward detecting the earliest changes of the disease in the brain, even before symptoms become apparent. It is generally believed that these brain changes can occur ten to twenty years before the symptoms. The hope of the scientists is that pharmaceutical or other type intervention could be instituted before the disease reaches some, yet undetermined, point of "no return" at which it is too late to halt or reverse the process through medication.

Without any medications that are effective long term, people with Alzheimer's and their loved ones are left to manage and struggle as best they can. The Alzheimer's Association provides excellent information about the nature of Alzheimer's disease and what research is going on and lists many helpful resources. Paid caregivers can provide some relief and support groups can offer comfort and practical advice. This is often not enough as the physical and emotional stress on families becomes overwhelming. When the condition of their loved one is such that families can no longer cope on their own, they may turn to institutional care which itself has significant problems. If all that were not enough, health care agencies and institutions, as well as pharmaceutical companies may be motivated by profit rather than the well-being of the patients.

What Can We Do?

Invest in Responsible Research and Beware of the "Hype"

Research efforts to elucidate the cause of Alzheimer's disease and to find a cure must be supported. It is only relatively recently that we have been hearing about more concerted efforts to stem the predicted tide of Alzheimer's disease by finding a cure. This is largely driven by the realization that the aging baby boomer generation and the expected explosive increase in the number of Alzheimer cases will have significant negative impact on the financial and social aspects of our country. You would think those facts alone would motivate our legislators. However, when well known comic and actor, Seth Rogen, went to Capitol Hill to share the personal story of his mother-in-law's five year decline from early onset AD, it was reported that 16 senators left before he gave his testimony and did not show up at the hearings at all.[1,2] So, while some legislators give "lip service" to wanting to help the cause, their actions, or more correctly, inaction, speak more loudly. We need to let them know that funding Alzheimer's research is a priority.

For the fiscal year 2015, Washington allocated on the order of $5.4 billion for research on cancer, $1.2 billion dollars for research on heart disease and $3 billion dollars for research on HIV/AIDS with only about $566 million going to research on Alzheimer's Disease. Ronald Peters, M.D. from the Mayo Clinic, Chair of an advisory committee for the National Alzheimer's Project, reported to Congress that $2 billion dollars annually would be necessary in order to meet the stated goals for Alzheimer's prevention and treatment.[3,4] The Federal budget for 2016 allocates an extra $350 million for Alzheimer's research. It will bring the total funding to $936 million and is a step in the right direction.

While we advocate for the funding of research, however, it is important that the studies be done in a responsible way and the results are properly validated. In the spring of 2015, a bill called the 21st Century Cures Act easily passed in the House of Representatives and on the surface sounds very promising. It has some provisions, such as annual increases in the budget for the National Institutes of Health (NIH), which certainly seem worthy, but other aspects of the proposed law are considered very controversial.[5-7] The law will try to speed up FDA approval for new drugs. This would be accomplished by using study designs, ways of analyzing data and determination of study endpoints by nontraditional methods. While it might make new medical drugs and devices available to patients more quickly, they may not work or they may have bad side effects that were not detected during the shortened trial period. With regard to Alzheimer's disease, an article published in the British Medical Journal noted that three Alzheimer's drugs formerly under investigation showed promising results in preliminary clinical trials based on analysis of the beta amyloid plaques or by tests of memory and cognition. However, two of the drugs for AD were found to be ineffective when larger, better designed trials were carried out and the other drug caused worsening in memory and an increased incidence of skin cancer.[6,7]

Most currently used Alzheimer's drugs are available in generic form, at relatively low cost, but their medical value is limited. Patients and their physicians would eagerly turn to a new and better medication, even if it was expensive. However, while there is much support for the new bill by pharmaceutical and medical device companies, others worry about the potential waste of money, and worse, the risk of harm to patients from a bad drug that was rushed to market.

Bring Alzheimer's Disease Into the Spotlight

Attention has been brought by the government to the negative consequences of Alzheimer's disease and the need to take action. In 2011, the National Alzheimer's Project Act, designed to fight the disease was signed into law. It called for further research and for better access to long term care. It provides for an updated plan to be submitted to Congress each year and calls for recommendations for improving the outcomes of people with Alzheimer's disease and for lowering costs to families. It created an Advisory Council on Alzheimer's Research, Care and Services.[8] In 2014 and 2015, there was also an impetus through various budget proposals and laws to increase funding for research.[9]

Nowadays, largely through the efforts of the Alzheimer's Association, there is also more awareness that Alzheimer's is a serious progressive illness and the stigma associated with the disease is beginning to lessen. The act of bringing the problem of dementia out of the dark and into the spotlight has also been aided by the famous people in all walks of life who have bravely shared their stories with the public. President Ronald Reagan was one of the first of these when he revealed his diagnosis in a letter to the country. Charlton Heston and his wife openly discussed his diagnosis of Alzheimer's disease in 2002 in a television interview. Lisa Gibbons talked about her mother and grandmother who had the disease. Maria Shriver has spoken about her family's experience when her father, Sargeant Shriver, developed Alzheimer's. She has been a strong advocate for research and for the caretakers of individuals with Alzheimer's disease. More recently, Glen Campbell brought attention to AD with his musical tour around the country and somewhat later we learned that he was no longer able to perform. This is indeed a disease which has no respect for power, genius, fame or fortune. We may feel that if these brilliant and talented people could not survive, what hope is there for us? There **is** hope, however, and we should never give up!

Encourage Social Interaction

Unfortunately, people with dementia have sometimes been seen as a source of shame or embarrassment. It is good to know, therefore, that the establishment of dementia friendly communities has been occurring in parts of Europe, including the United Kingdom, Scotland and Ireland, and has also begun in a small way in the United States. For example, selected businesses in Watertown, Wisconsin display small decals indicating that the store's employees have training in recognizing and helping people with dementia. There is a Memory Café for people with dementia and their caretakers who meet monthly. It serves as both a support and social group.[10,11]

Social interaction has been recognized as a factor in helping to preserve cognition.[12] In Minnesota, the American Association of Retired People (AARP) joined with over 50 groups to advise communities on dealing with the increasing number of patients with dementia. These types of endeavors help to relieve the isolation and stigma facing people with dementia and their loved ones. According to the 2012 World Alzheimer Report, roughly one-fourth of people with dementia are aware of the stigma attached to the diagnosis and many report being excluded from activities in which they would ordinarily partake.[13] With all the focus on the medical aspects and financial ramifications of Alzheimer's disease, we should not forget the plight of the individual patient himself or herself.

Help the Caregivers

Most of the long term care of dementia patients in the United States takes place in their homes and is provided by unpaid caregivers, usually family and friends. Most of the caregivers are women. It is frequently reported that the hours of care that they provide is equivalent to a part time job although if you have been involved in this type of care giving you know it can amount to virtually round the clock care. Many caregivers who have no family support and who cannot afford outside help, continue in this stressful manner for years without any backup or time off. These caregivers may have physical problems of their own which they don't have time to address. They are forced into a state of social isolation and are more vulnerable to depression. Add to this the financial burdens and lost wages. Employers can help in this regard by offering a more flexible work schedule.

Real Life Learning for Doctors-To-Be

Dementia patients are paired with doctors in training for a year as part of a "buddy" program which was pioneered at Northwestern University in Chicago. This novel teaching experience has since been instituted by a few other medical schools. It gives medical students a real life one-on-one perspective of the human impact of Alzheimer's disease in a way that cannot be learned in classrooms or from textbooks. It is also an avenue by which the dementia patients, at least in the earlier stages of their disease, can participate in a meaningful and beneficial social interaction[14,15] Anything that takes patients with dementia out of the shadows and keeps AD in the mindset of the medical community is a good thing.

Try to Give the Alzheimer's Patient a Better Quality of Life

As a reaction to the deficiencies in long term care institutions, some Americans and western Europeans have actually brought their loved ones with Alzheimer's disease to other countries for their care. They have found that the care in these places is much cheaper and also much less institutional in nature

In one of these communities in Thailand called Baan Kamlangchay ("Home for Care from the Heart"), patients live in individual houses and are taken to local markets and restaurants. Each patient has personal round-the-clock care provided by three caretakers working in shifts. The residents participate in outdoor activities regularly. They are given freedom to move about, although the caretakers make sure that they are safe. The monthly cost is $3,800. American patients can receive care in the Philippines for $1,500 to $3,500 monthly.[16]

This trend is not without controversy, however. Some experts say that uprooting people with Alzheimer's disease could have the harmful effect of adding to their anxiety and that they should remain in their familiar environment as long as possible where friends and family can visit them. Others believe that the quality, rather than the location, of the care given is more important and that over time even those in advanced stages of dementia can adjust. In any case, sending their loved ones abroad is probably pretty extreme for most families but it speaks to the deficiencies in our own system of long term care and the level of desperation of some caretaker relatives.

That brings us to the question of how we can change practices in nursing homes so that those people in the later stages of Alzheimer's disease who find themselves in their care have a safer, kinder environment in which to live and a better quality of life. Some professionals in the fields of gerontology and dementia have already been recommending improvements in the quality of care evaluations of long term care facilities. For example, the monitoring quality in nursing homes and other type care homes should include *quality of life* issues for the residents, in addition to safety and other important but routine inspections. Patients should be encouraged to have as much autonomy as is practically possible for their condition. People with dementia have unique problems and needs and the staff members that take care of them need to be much better trained to address these issues.

Nursing homes and facilities that care for late stage patients with Alzheimer's disease should be fully staffed and the staffing should be adequate to provide the optimum level of care not only during the day but also at night. There should be appropriate financial remuneration for the staff. The nurses and nursing assistants that take care of patients with dementia need to have more training and better understanding of their special problems and needs. For those people being taken care of at home, there should be recognition of the value of the unpaid caregiving work of family members with some type of remuneration, if possible. The deficiencies in long term care facilities are well known. Sadly, however, goals for improving conditions appear to have been just as difficult to attain as a medical cure for Alzheimer's. This may be due to the apathy of many and the greed of some. It would take a tremendous and concerted effort by those "in charge" and a change to some creative and constructive thinking. In the meantime, people should expect, and if necessary, demand from long term care facilities that their loved ones receive good care and be treated with dignity.

Cynthia Janus, M.D.

Why Should We Care?

One reason we should care is that Alzheimer's disease represents a growing financial burden and strain on our health care system. More important than the financial considerations, however, is the tremendous human toll that Alzheimer's takes. The world at large loses many talented and productive people to the disease. Individuals lose their parents, siblings and friends after watching them become gradually more incapacitated in every way. Think of all the children who are deprived of their grandparents, and of the elderly who lose their spouse of many decades during what should be their happy retirement years.

. If you are dealing with Alzheimer's on a personal basis as a caregiver, you know that it becomes the focus of your whole life. For everyone else, there are numerous events and situations in our personal lives or on the news that grab our attention. There are many charitable organizations and various causes that reach out to us for contributions. There are many groups of people who desperately need help, such as victims of natural disasters, cancer patients, victims of all types of abuse, and our veterans. These are extremely important causes and are certainly worthy of our attention and contributions.

I have wondered why the needs of people with Alzheimer's disease have not called out to us in as compelling a manner. There has been comparatively little in the way of wristbands, "walks for the cure" or awareness campaigns comparable in scope to those that we have seen successfully utilized for breast cancer, AIDS and other causes. Is it because the victims of AD, especially late in the course of the disease, cannot speak out for themselves to plead their case? Is it possible that we think of Alzheimer's disease as merely a memory disorder and are not fully aware of its life destructive nature? Do we unconsciously feel that individuals with Alzheimer's disease are (usually) elderly, so maybe it doesn't matter as much? Perhaps we believe there's really not much we can do, so why bother? Maybe, we can't really identify with the cause (the "it can't happen to me" mentality). I hope not because, unless later proven otherwise, **_WE ARE ALL VULNERABLE_**.

I believe that the medical breakthrough that will put an end to the suffering that Alzheimer's disease inflicts will one day come to pass.

Hopefully, it will happen soon. However, it will likely take a long while and come slowly in the form of multiple small discoveries that build on each other. Then, it will take considerably more time to establish that any new medication is truly effective and without serious side effects. In the meantime, while we are waiting, we should not accept the status quo but instead make every effort to make life better for those people with Alzheimer's disease. There is too much at stake.

ABOUT THE AUTHOR

Cynthia Janus, M.D. was Associate Professor of Radiology at the Mt. Sinai Medical Center in New York City and at the University of Charlottesville in Virginia, and has written extensively on medical topics. She also co-authored *The Janus Report on Sexual Behavior.* Dr. Janus currently lives in Tampa, Florida where she maintains a keen interest in current medical and sociological issues.

ENDNOTES

Chapter 1

1. Helpguide.Org. *Understanding Dementia: Signs, Symptoms, Types and Treatment.* http://www.helpguide.org/elder/alzheimers_dementias_type.htm
2. Media centre news release, 4-11-12. World Health Organization. *Dementia cases set to triple by 2050 but still largely ignored.* http://www.who.int/mediacentre/news/releases/2012/dementia_2012041/en/
3. Jin J. *Alzheimer Disease.* JAMA Patient Page/Neurology JAMA 313 (14) : 1480, 4-14-15
4. The National Cell Repository for Alzheimer's Disease. *The Brain and Alzheimer's Disease, NCRAD>The Importance of Autopsy>*The Brain and AD. http://nerad.iu.edu./Autopsy/brainAD.asp
5. Malone DC, McLaughlin TP, Wahl PM, et al. *Burden of Alzheimer's disease and association with negative health outcomes.* Am J Manag Care 15(8):481-488, 2009.
6. Alz.org/alzheimer's association. *Mild Cognitive Impairment.* http://www.alz.org/dementia/mild-cognitive-impairment-mei.asp
7. Weintraub, K. *Estimate: 135 million worldwide with dementia by 2050*; Special for USA_Today, 12-5-13 https://twitter.com/intent/tweet?url=htttp://usat.ly/1bhQ3Ep+text=Estiimate:%20135%20million.
8. Alz.org/alzheimer's association. Latest Facts and Figures Report/Alzheimer's Association. *2015 Alzheimer's Disease Facts and Figures.* http:/www.alz.org/facts/
9. Belluck, P. *Dementia Care Cost is Projected to Double by 2040.* New York Times/Health, 4-3-13 www.nytimes.com/2013/04/04/health/dementia/care-costs-are-soaring-study-finds.html?_r=0
10. Seegert, L. *RAND Study: More LTSS for Alzheimer's patients, caregivers needed.* Association of Health Care Journalists, 6-26-14 http://healthjournalism.org/blog/2014/06/rand-study-more-ltss-for-alzheimers-patients-caregivers-needed/
11. Centers for Disease Control and Prevention._*Leading Causes of Death.* www.cdc.gov/nchs/fastats/leading-causes-of-death.htm
12. Bahrampour, T. *New Study Ranks Alzheimer's as third-leading cause of death, after heart disease and cancer.* The Washington Post, 3-5-14 http://www.

washingtonpost.com/local/new-study-ranks-alzheimers-as-third-leading-cause-of-death-after-heart-disease-and-c ancer/2014.03/05/8097a452-a43a-1...

13   James BD, Leurgans SE, Hebert LE, et al. *Contribution of Alzheimer disease to mortality in the United States.* http://www.neurology.org/content/early/WNL.0000000000000240

14   The Editorial Board of the New York Times. *High Mortality From Alzheimer's Disease.* Editorial, New York Times, 3-12-14. http://www.nytimes.com2014//0313/opinion/high-mortality-from-alzheimer'sdisease.html?ref=opinion&_r=1

15   Goldschmidt, D. *Report: Alzheimer's far more likely than breast cancer in women over 60.* CNN/Health, 3-19-14 http://www.cnn.com/2014/03/19/health/women-alzheimers/

16   RAND Corp Press Room News Release. *Cost of Dementia Tops $159* Billion Annually in the United States.* RAND Corp, 4-3-13. http://www.rand.org>Press room>News Releases>2013

17   Lloyd, J. *As Alzheimer's rate soars, concern rises over costs.* USA Today, 2-6-13. http://www.usatoday.com/story/news/nation/2013/02/14alzheimers-cost-medicare-health/1917513/

18   Serna, J. *Alzheimer's cases, and costs, projected to swell.* Los Angeles Times/Science, 2-7-2013. http://www.latimes.com/news/science/la-sci-alzheimers-disease-boom-20130207,0,6234076.story

19   Preidt, R. *Alzheimer's "Epidemic Straining Caregiver, Community Resources: Report.* HealthDay, 9-19-13. http://www.consumer.healthday.com/cognitive-and-neurological-health-information-26/alzheimer-s-news-20/briefs-emb-9-19-3pmet-alzheimer-s-report-kol-release-batch-9...

20   Healy, M. *With millions more expected to develop Alzheimer's, more research funding demanded.* LA Times, 7-20-15. http://www.latimes.com/science/sciencenow/la-sci-sn-baby-boomers-alzheimers-20150700-story.html.

Chapter 2

1   Kennard, C. *Did Dr. Alzheimer discover Alzheimer's Disease?* About.com. Alzheimer's/ Dementia, 7-31-06. Alzheimers.about.com/cs/caregiver/a/Alois_Alzheimer.htm

2   Alzheimer's Disease Education and Referral Center/National institute on Aging. *Alzheimer's Disease: Unraveling the Mystery.* http://www.nia.nih.gov/alzheimers/publication/part-2-what-happens-brain-ad/hallmarks-ad

3   National Institute of Health/National Institute on Aging. Alzheimer's Disease Education and Referral Center. Alzheimer's Disease Medications Fact Sheet, 7/ 2010, page last updated 12-13-11. http://www.nia.nih.gov/alzheimers/publication/alzheimers-disease-medications-fact-sheet

4   Mayo Clinic Staff. *Alzheimer's : Drugs help manage symptoms*. Mayo Clinic: Diseases and Conditions/ Alzheimer's Disease. 7-11-14. http://www.mayoclinic.org/diseases-conditions/alzheimers-disease/in-depth/alzheimers/art-20048103
5   Lopez OL, Becker JT, Wisniewski S, et al: *Cholinesterase inhibitor treatment alters the natural history of Alzheimer's disease*. J Neurosurg Psychiatry 72: 310-314, 2002 doi:10.1136/jnp.72.3.310 http://jnnp.bmj.com/content/72/3/310
6   Doody RS, Ferris SH, Salloway S, et al. Abstract: Donepezil treatment of patients with MCI: a 48-week randomized, placebo-controlled trial. Full article published in Neurology 72(18): 1555-1561, 2009. http://www.ncbi.nim.nih.gov/pubmed/19176895
7   Sano M, Ernesto C, Thomas RG, et al. Abstract: *A controlled trial of selegine, alpha-tocopherol, or both as treatment for Alzheimer's disease. The Alzheimer's Disease Cooperative Study*. Full article published in N Engl J Med, 336(17):1216-22, 4-24-97 http://www.ncbi.nih.gov/pubmed/9110909
8   Alz.org/Alzheimer's Association. *Medications for Memory Loss* http://www.alz.org/alzheimers_disease_standard-prescriptions.asp
9   Thomas, K. *Drug Dosage Was Approved Despite Warning*. The New York Times/Business Day, 3-22-12. www.nytimes.com/2012/03/23/business/drug-dosage-was-approved-despite-warning.html?_r=1
10 JAMA/Health Agencies Update. *Challenge to Alzheimer Drug*. JAMA 38(24): 2557, 12-26-12
11 Schoenberg T. *FDA Sued by Consumer Group Over Alzheimer Drug Dosage*. Bloomberg News, 9-6-12 http://bloomberg.com/news/2012-09-05/fda-sued-by-consumer-group-over-alzheimer-drug-dosage.html
12 Silverman, E. *Actavis Must Keep Selling Old Version of Alzheimer's Drug Namenda, Court Rules*. The Wall Street Journal. Updated 5-22-15. http://www.wsj.com/articles/actavis-must-keep-selling-old-version-of-alzheimers-drug-namenda-court-rules-1432326861
13 Tjia J, Briesacher BA, Peterson D, et al. Abstract: *Use of Medications of questionable benefit in advanced dementia*. Full article in JAMA Intern Med 174(11): 1763-71, 2014 http://www.ncbi.nim.nih.gov/pubmed/25201279
14 Healy, M. *Dementia patients continue to get medications with little, no benefit*. LA Times. 9-9-14 http://www.latimes.com/science/sciencenow/la-sci-sn-dementia-medications-benefit-20140908-story.html
15 Kunkle F. *Alzheimer's costs could soar to $1 trillion a year by 2050, report says*. TheWashington Post, 2-2-15
16 Tayeb HO, Murray ED, Price BH, et al. Abstract: *Bapineuzumab for Alzheimer's Disease: Is the "amyloid cascade hypothesis" still alive?* Full article in Expert Opin Bio Ther. 13(7): 1075-84, 2013 http://www.ncbi.nim.nih.gov/pubmed/23574434

17. Cortez, MF. *Pfizer-J & J Alzheimer's Drug Shows Promise for Early Use.* Bloomberg, 9=11=12. http://www.bloomberg.com/news/2012-09-11/pfizer-j-j-alzheimer-s-drug-shows-promise-for-early-use.html
18. Thompson, D. *2 Alzheimer's Drugs Found Ineffective in Clinical Trials.* HealthDay 1-22-14 http://consumer.healthday.com/cognitive-health-information-s-news-20/alzheimer-s-drugs-684115.html
19. Alzheimer Forum, Gwyneth Dickey Zakaib. Jan 22, 2014. *Phase 3 Trial Data for Solanezumab and Bapineuzumab Release* http://www.alzforum.org/news/research-news/phase3-trial-data-solanezumab-and-bapineuzumab-released
20. Kolata, G. *Three Drugs to Be Tested to Stave Off Alzheimer's.* New York Times/ Health. Oct 10, 2012. www.nytimes.com/2012/10/11/health/alzheimers-prevention-studies-to-test-three-drugs.html?_r=0
21. Sussex Drug Discovery Group. *Amyloid in Alzheimer's Disease-The End of the Beginning or the Beginning of the End?* 1-7-13. http://sussexdrugdiscovery.wordpress.com/2013/01/07/amyloid-in-alzheimers-disease-the-end-of-the-beginnning-or-the-beginning-of-the-end/
22. Pierson, R. *Biogen plans late-stage Alzheimer's trial, shares rise.* Reuters, 12-2-14. http://www.reuters.com/article/2014/12/02/us-biogen-idec-alzheimers-idUSKCNOJG1JX20141202
23. Cummings JL, Morstorf T, Zhong K. Abstract: *Alzheimer's disease drug development pipeline: few candidates, frequent failures.* Article in Alzheimers Res Ther 6(4): 37,2014 PubMed-NCBI. http://www.ncbi.nlm.nih.gov/pubmed/25024750
24. Harasim P. *Alzheimer's drugs not working, according to Ruvo Center study.* Las Vegas Review Journal, 7-2-14. http://www.reviewjournal.com/life/health/alzheimer-s-drugs-not-working-according-ruvo-center-study
25. Norton A. *Alzheimer's-Linked Brain Plaques May Arise Decades Before Symptoms.* HealthDay, 5-19-15. http://consumer.healthday.com/senior-citizen-information-31/misc-aging-news-10/alzheimer-s-linked-brain-plaques-may-arise-decades-before-symptoms-69...
26. Snitz BE, O'Meara ES, Carlson MC, et al. Abstract: *Ginkgo Biloba for preventing cognitive decline in older adults: a randomized trial.* Full article in JAMA 302(24):2663-70, 2009 Http://www.ncbi.hlm.nih.gov/pubmed/20040554
27. Reinberg S. *Ginkgo Won't Prevent Alzheimer's, Study Finds.* HealthDay, 9-5-12. Consumer.healthday.com/article.asp?AID-668378
28. Rogers MB. *Gammagard Misses Endpoints in Phase 3 Trial.* AlzForum, 5-8-13, http://www.alzforum.org/news/research-news/gammagardtm-misses-endpoints-phase-3-trial
29. Jeffrey S. *GAP: IVIG Negative in Alzheimer's, But Some Hints of Benefit.* Medscape, 7-17-13. http://wwwomedscape.com/viewarticle/807970

30. Yan R, Vassar R. Abstract: *Targeting the B secretase BACE 1 for Alzheimer's disease therapy*. Full article in Lancet Neurol 13(3): 319-29, 2014, HHS Public Access http://www.ncbi.nlm.nih.gov/pmc/articles/PMC4086426/

31. Yan X-X, Ma C, Gai W-P, et al. Abstract : *Can BACE1 Inhibition Mitigate Early Axonal Pathology in Neurological Diseases*. HHS Public Access. Article in J Alzheimers Dis 38(4):705-18, 2010 Httap://www.ncbi.nim.nih.gov/pmc/articles/pmc3995167.

32. Neergaard L. *Testing brain pacemakers to zap Alzheimer's damage*. Associated Press, 1-21-13. Www.google.com/hostednews/ap/article/ALeqM5jRrzaGqR8FpKcuNRhZ48sfZFRQTA?docId=64b276ac9c1241f5b7d99451378ce3b4

33. Sankar T, Chakravarty MM, Bescos A. Abstract: *Deep Brain Stimulation Influences Brain Structure in Alzheimer's Disease*. Article in Brain Stim 8(3): 645-54, 2015

34. Friedrich, MJ. *Studies Suggest potential Approaches for Early Detection of Alzheimer Disease*. JAMA 309(1):18, 1-2-13

35. Preidt R. *Leaks in Brain May Contribute to Dementia*. HealthDay, 1-21-15. http://consumer.healthday.com/senior-citizen-information-31/misc-aging-news-10/leaks-in-brain-may-contribute-to-dementia-695690.htm

36. Whalen. *Ultrasound Shows Promise in Mice for Treating Alzheimer's*. The Wall Street Journal, 3-11-15. http://wsj.com/articles/ultrasound-treatment-helps-improve-memory-in-mice-1426096802

37. Thompson D. *Is Tau the 'How' Behind Alzheier's?* HealthDay, 10-31-14 http://consumer.healthday.com/cognitive-health-information-26/alzheimer-s-news-20/is-tau-the-how-behind-alzheimer-s-693274.html

38. Preidt R. *Researchers Pinpoint Possible Protein Culprit Behind Alzheimer's*. HealthDay, 3 24 15. http://consumer.healthday.com/cognitive-health-information-26/alheimer-s-news-20/abnormal-tau-protiens-the-driving-force-behind-alzheimer-s-study-claim...

39. Hirschler B, Copley C. *Update 2-J&J strikes Alzheimer's research deal with Swiss firm AC Immune*. Reuters, 1-2-15. http://www.reuters.com/article/2015/01/12/ac-immune-jj-idUSL6NOUR0MP20150112

40. Kan MJ, Lee JE, Wilson JG, et al. Abstract : *Arginine Deprivation and Immune Suppression in a Mouse Model of Alzheimer's Disease*. Article in The Journal of Neuroscience. April 15, 2015, 35(15):5969-82, 4-15-15. http://www.jneurosci.org/content/35/15/5969.short?sid=8f734ac8-e053-4f379dad-65fa3777809d

41. Priedt, R. *Mouse Study Suggests Immune Disorder May Play Role in Alzheimer's*. U.S. News & World Report/Health Day, 4-14-15 http://health.usnews.com/health-news./articles/2015/04/14/mouse-study-suggests-immune-disorder-may-play-role-in-alzheimers

42. Cleveland Clinic: *Alzheimers Disease and Down Syndrome*. 2011 http://my.clevelandclinic.org/health/diseases_conditions/hic_Alzheimers_and_Dementia_Overview/hic_Alzheimers_Disease_and-Down-Syndrome
43. Mayo Clinic. Diseases and Conditions/Alzheimer's Disease. *Early-onset Alzheimer's; Genetic Testing*. http://www.mayoclinic.org/diseases-conditions/Alzheimer's-disease/in-depth/alzheimers-genes/art-20046552?pg=2
44. Mayo Clinic Staff. *Alzheimer's genes: Are you at risk? Most common late-onset Alzheimer's gene*. Mayo Clinic/ Diseases and Conditions: Alzheimer's Disease 2013 http://www.mayoclinic.org/diseases-conditions/alzheimers-disease/in-depth/alzheimers-genes/art-20046552
45. Woemer A. New *Alzheimer's Discovery Could Hold Key to Preventative Treatments for High-Risk Patients*. Fox News, 10-22-13. http://www.foxnews.com/health/2013/10/22/new-alzheimers-discover-could-hold-key-to-preventative-treatments-for-high/?intcmp=obnetwork
46. Lloyd J. *New Gene Holds Promise for Understanding Alzheimers*. USA Today, 11-12-12. http://www.usatoday.com/story/news/nation/2012/11/14/alzheimers-treatment-cure-prevention/1705239/
47. Alzheimer's Disease Education and Referral Center. Identifying the Genetics of Alzheimer's Disease. 2013-2014 Alzheimer's Disease Progress Report. http://www.nia.nih.gov/alzheimers/publication/2013-2014-alzheimers-disease-progress-report/identifying-genetics-alzheimers
48. *Drug restores brain function and memory in early Alzheimer's disease*. http://medicalxpress.com/news/2015-03-drug-brain-function-memory-early.html. 3-11-15
49. Mozes A. *Experimental Alzheimer's Drug Shows Promise, Study Finds*. HealthDay, 3-11-13. Consumer.healthday.com/Article.asp?AID=674306
50. Dysken M, Sano M, Asthana S, et al. *Effect of Vitamin E and Memantine on Functional Decline in Alzheimer Disease*. JAMA 311(1):33-44, 1-1-14
51. Chakravorty S. *Ten Rivals Join With NIH to Search For New Drugs – WSJ*. Reuters, 2-4-14. http://www.reuters.com/article/2014/02/04/us-nih-drugspact-idUSBREA130BI20140204
52. Mattson N, Zetterberg H, Hansson O, et al. Abstract : *CSF Biomarkers and Incipient Alzheimer Disease in Patients With Mild Cognitive Impairment*. Full article in JAMA 302(4) 385-93, 2009. http://jama.jamanetwork.com/article.aspx?articleid=184311
53. Buchhave P, Minthon L, Zetterberg H, et al. Abstract : *Cerebrospinal fluid levels of B-amyloid 1-42, but not tau, are fully changed already 5 to 10 years before the onset of Alzheimer dementia*. Full article in Arch Gen Psychiatry 69(1):98-106,2012. http://www.ncbi.nlm.nih.gov/pubmed/22213792
54. Laske C, Stransky E, Leyhe T, et al. *Stage-dependent BDNF serum concentrations in Alzheimer's disease*. J Neural Transm 113: 1217-24, 2006

55. Richards BJ. *BDNF Prevents and Reverses Alzheimer's Disease.* Health & Wellness News, 2-11-09. http://www.wellnessresources.com/health/articles/bdnf_prevents_and_reverses_alzheimers_disease/
56. Ventriglia M, Zanardini R, Bonomini C, et al. Abstract: Serum Brain-Derived Neurotrophic Factor Levels in Different *Neurological Diseases.* Article in BioMed Researh International. Vol 2013(2013), Aritcle ID901082, 7pages http://www.hindawi.com/journals/bmri/2013/901982
57. Lim JY, Reighard CP, Crowther DC. Abstract: *The pro-domains of neurotrophins, including BDNF, are linked to Alzheimer's disease through a toxic synergy with AB.* http://hmg.oxfordjournals.org/content/early/2015/04/22hmg.ddv130.abstract
58. Cohen E. *Blood Test Predicts Alzheimer's Disease.* CNN Health, 3-9-14. http://www.cnn.com/2014/03/09/health/alzheimers-blood-test/
59. Mayer, K. *Smell and Eye Tests May Permit Early Detection of Alzheimer's.* Genetic Engineering and Biotechnology News. http://www.genegnews.com/gen-news-highlights/smell-and-eye-tests-may-permit-early-detection-of-alzheimer-s/81250099/
60. Fischer, K. *Imaging Test Might Identify Biomarker of Alzheimer's Disease: Scientists say deteriorating white matter in the brain could be an early indicator of Alzheimer's.* Healthline News, 5-27-15.
61. Preidt R. *White matter damage in brain may help spot early Alzheimer's.* HealthDay, 5-27-15. http://consumer.healthday.com/cognitive-health-information-26/a;zheimer-s-news-20/white-matter-damage-in-brain-may-help-identify-early-alzheimer-s-6996
62. Healy M. *Radiologists use MRIs to find biomarker for Alzheimer's disease.* LA Times, 10-7-14. http://www.latimes.com/science/sciencenow/la-sci-sn-alzheimers-brain-biomarker-radiology-20141007-story.html
63. Jansen WJ, Ossenkoppele R, Knol DL, et al. *Prevalence of Cerebral Amyloid Pathology in Persons Without Dementia: A Meta-Analysis.* JAMA 313(19): 1924-38, 5-19-15
64. Ossenkoppele R, Janssen WJ, Rabinovici GD, et al. *Prevalence of Amyloid PET Postivity in Dementia Syndromes: A Meta-Analysis.* JAMA 313(19):1939-49, 5-19-15

Chapter 3

1. Ramachandran TS. *Alzheimer Disease Imaging.* Medscape: Drugs, Diseases and Procedures, Updated 2-15-12. medicine.medscape.com/article/336281-overview#showall
2. Gale, J. *Alzheimer's Seen on Scans Decades Before Dementia, StudyShows.* Bloomberg, 3-7-13. www.bloomberg.com/news/2013-03-08/alzheimer-s-seen-on-scans-decades-before-dementia-study-shows.html

3. Cortez, MF. *Insurers Urged to Cover Brain Imaging for Alzheimer's*. Bloomberg, 1-28-13. www.bloomberg.com/news/2013-01-28/Insurers-urged-to-cover-brain-imaging-for-alzheimer-s.html

4. Jack CR, Wiste HJ, Vemuri P, et al. *Brain Beta-amyloid Measures ad Magnetic Resonance Imaging Atrophy both Predict Time-to-progression from Mild Cognitive Impairment to Alzheimer's Disease*. Published in Brain 133(11)3336-48, 2010. http://www.medscape.com/viewarticle/732132_4

5. Mitsis EM, Bender HA, Kostakoglu L, et al. Abstract: *A consecutive case series experience with [^{18}F] florbetapir PET imaging in an urban dementia center: impact on quality of life, decision making and disposition*. http://www.ncbi.nig.gov/pmc/articles/PMC3913628/

6. Yeager, D. *PET May Predict Future Cognitive Decline*. Radiology Today: Molecular Imaging News, P.8, May 2014

7. Mitka, M. *PET Imaging for Alzheimer Disease: Are Its Benefits Worth the Cost?* Medical News and Perspectives. JAMA 309(11):1099-1100, 3-20-13

8. Pittman D. *Medicare Panel Pans Alzheimer's Test*. MedPage Today, 1-30-13. http://www.medpage.today.com/Neurology/AlzheimersDisease/37116

9. Tadena N, *Eli Lilly "Disappointed" in CMS Decision on Amyvid*. Wall Street Journal, 9-30-13. onlinewsj.com/article/BT-CO-20130930-704278.html

10. Orenstein, BW. *Amyloid Imaging: Societies Produce Guidelines for Amyvid PET Use*. Radiology Today, April 2013. www.radiologytoday.net

11. *$100M IDEAS: CMS Blesses Study to Evaluate Amyloid Scans in Clinical Practice*. ALZFORUM: Networking for a Cure, 4-16-15. http://www.alzforum.org/news/community-news/100m-ideas-cms-blesses-study-evaluate-amyloid-scans-clinical-practice

12. Fiore K. *Amyloid Imaging May Shift Alzheimer Management*. MedPage Today; Meeting Coverage, 7-22-15. http://www.medpagetoday.com/MeetingCoverage/AAIC?52719

13. *Imaging Alzheimer's: Early Data Show FDG-PET Scan With Early Treatment Improves Outcomes*. Source: UCLA. Radiology Today, October 2013. www.radiologytoday.net

14. Woock K, *Coverage for Some*. Imaging Economics: Regulatory Watch. Nov/Dec 2013 Imagingeconomics.com

15. Le Couteur DG, Brayne C. *Should Family Physicians Routinely Screen Patients for Cognitive Impairment? No: Screening Has Been Inappropriately Urged Despite Absence of Evidence*. Published in Am Fam Physician 89(11):864-5, 6-1-14. http://wwww.aafp.org/afp/2014/0601/p864.html

16. Aschwanden, C. *Why you may want to avoid a dementia test*. The Washington Post/Health and Science. www.washingtonpost.com/national/health-science/why-you-way-want-to-avoid-a-dementia-test/2013/12/13/6377bee6-6071-11e3-bf45-61f69f54fc5f_story.html

17. Beck M. *Alzheimer's Patients Aren't Always Told They Have Alzheimer's.* The Wall Street Journal/Health and Wellness. Updated 3-23-15. http://www.wsj.com/articles/alzheimers-patients-aren't-always-told-they-have-alzheimers-1427169935
18. Kaplan DA. Q&A: *The Prevalence of Overdiagnosis in Alzheimer's Disease.* Consultant Live. Q&A with Jacob Dubroff, MD., PhD., 7-27-15. http://www.consultantlive.com/petmr/qa-prevalence-overidagnosis-alzheimers-disease
19. Zarembo A. *FDA forces UCLA researchers to stop touting experimental dementia scan.* LA Times, 4-10-15. http://www.latimes.com/science/sciencenow/la-sci-sn-brain-scan-warning-taumark-20150410-story.html
20. Reinberg S. *Jury Still Out on Routine Dementia Screening for Seniors.* Health Day, 3-24-14. http://www.consumer.healthday.com/cognitive-health-information-26/alzheimer-s-news-20froutine-screening-for-dementia-not-beneficial-report-686099.html

Chapter 4

1. Healy M. *Cholesterol and Alzheimer's disease link strengthens in study.* Los Angeles Times/Science, 12-30-13. www.latimes.com/science/sciencenow/la-sci-cholesterol-alzheimers-link-20131230,0,4639284.story
2. Park A. *Statin Drugs Linked to Lower Risk of Cognitive Decline.* 10-1-13. Healthland.time.com/2013/10/01/statin-drugs-linked-to-lower-risk-of-cognitive-decline/
3. JAMA Neurology/ Network Abstracts. *Associations Between Cerebral Small-Vessel Disease and Alzheimer Disease Pathology as Measured by Cerebrospinal Fluid Biomarkers.* JAMA 312(10), 9-10-14
4. Lloyd, J. *Blood pressure drugs may also reduce dementia risk.* USA Today, 1-7-13. www.usatoday.com/story/nation/2013/01/07/hypertension-dementia-blood pressure/1810369/
5. Healy M. Strict blood pressure control won't stem mental decline, study says. Los Angeles Times/Science, 3-3-14. http://www.latimes.com/science/sciencenow/la-sci-sn-hypertension-mental-decline-20140303,0,2827204.story
6. Willette AA, Bendlin BB, Starks EJ, et al. Abstract: Association of Insulin Resistance with Cerebral Glucose Uptake in Late Middle-Aged Adults at Risk for Alzheimer Disease. Article in JAMA Neurol. Published online, 7-27-15. Doi.10.10001/jamaneurol.2015,0613. http://archneur.jamanetwork.com/article.aspc?articleid=2398420
7. Kitamura M. *Fasting at Least Twice a Week Seen as Alzheimer's Hedge.* Bloomberg, 10-28-13. www.bloomberg.com/news/2013-10-29/fasting-at-least-twice-a-week-seen-as-alzheimer-s-hedge.html

8 Johansson L, Guo X, Duberstein PR, et al. *Midlife personality and risk of Alzheimer disease and distress.* Article published in Neurology 83(17): 1538-44, 10-21-14. http://www.neurology.org/context/83/17/1538

9 Norton A. *Apathy Might Signal Brain Shrinkage in Old Age: Study.* Healt Day News, 4-26-14. http://consumer.healthday.com/senior-citizen-information-31/misc-aging-news-10/apathy-may-be-sign-of-brain-shrinkage-in-old-age-686892.html

10 Doyle, K. *Depression linked to faster cognitive decline in old age.* Reuters, 7-30-14. http://www.reuters.com/article/2014/07/30/us-dementia-depression-id USKBN0FZ2CD20140730

11 Molecular Imaging News. *Depression in the Elderly Linked to Alzheimer's Risk.* Radiology Today, P. 26. July 2014

12 Healy M. *Preventing Alzheimer's disease – with an antidepressant.* Los Angeles Times/Science Now, 5-14-14. http://www.latimes.com/science/sciencenowla-sci-sn-alzheimers-antidepressant-20140513-story.html

13 Fox News. *Countries with more wealth, better hygiene have higher Alzheimer's risk.* Fox news.com, 9-6-13. www.foxnews.com/health/2013/09/06/countries-with-more-wealth-better-hygiene-have-higher-alzheimers-risk/

14 Weintraub K. *Brain injury in veterans tied to Alzheimer's risk.* USA Today. 6-26-14. http://usat.ly/1v7KY8a

15 Healy M. *After an ICU stay, cognitive loss is common, study says.* LA Times/Science. 10-2-13. www.latimes.com/science/sciencenow/la-sci-intensive-care-cognitive-loss-20131002,0,2845616.story

16 JAMA: Health Agencies Update. *Sleep May Help Remove Harmful Molecules From the Brain.* JAMA 32(20):2140, 11-27-13

17 Luscombe B. *Your Brain Cells Shrink While You Sleep (And That's a Good Thing)* Healthland.time.com/2013/10/17/your-brain-cells-shrink-while-you-sleep-and-that's-a-good-thing/

18 Neergaard L. *Studies: Better Sleep May Be Important for Alzheimer's Risk.* News From the Associated Press, 7-20-15. http://hosted.ap.org/dynamic/stories/U/US_MED_Alzheimers_SLEEP/SITE=AP&SECTION=HOME&TEMPLATE=DEFAULT

19 Marcus MB. *High Estrogen Levels Plus Diabetes May Boost Dementia Risk.* HealthDay, 1-29-14. http://consumer.healthday.com/senior-citizen-information-31/misc-aging-news-10/high-estrogen-levels-plus-diabetes-may-boost-dementia-risk-684324.html

20 Shumaker SA, Legault C, Kuller L, et al. Abstract: *Conjugated equine estrogens and incidence of probable dementia and mild cognitive impairment in postmenopausal women: Women's Health Initiative Memory Study.* Article published in JAMA 291(24): 2947-58, 6-23-04

21. Kunkle, F. *Why do more women get Alzheimer's? Research points to genetics, other factors.* The Washington Post. http://www.wpost.com/local/with-women-perhaps-facing-higher-risk-of-alzheimers-female-scientists-unite/2014/09/03/2aa0506c-28ab-11e4-8593-da634b334390...

22. Marcus MB. *Could Infections Harm Memory in Older Adults?* HealthDay, 2-13-14. http://consumer.healthday.com/senior-citizen-information-31misc-aging-news-10/could-certain-infections-harm-memory-in-older-adults-684691.html

23. Richardson JR, Roy A, Shalat SL, et al. Abstract: *Elevated Serum Pesticide Levels and Risk for Alzheimer Disease.* Article in JAMA Neurol. Published online January 27, 2014. doi:10,1001/jamaneurol.2013.6030 http://archneur.jamaneetwork/com/article.aspx?articleid=1816015

24. Healy M. *Drugs used for anxiety, sleep are linked to Alzheimer's disease in older people.* LA Times, 9-9-14. http://www.latimes.com/science/la-sci-sn-anxiety-drug-alzheimers-20140909-story.html

25. Roberts M, *Dementia "linked" to common over-the-counter drugs.* BBC News/Health, 1-27-15. http://www.bbc.com/news/health-30988643

26. Gastaldo E. *Benadryl may increase risk of Alzheimer's, study says.* http://www.fox/news.com/health/2015/01/27/study-benadryl-may-up-risk-alzheimer/?intcmp=ob_homepage_health&intcmp=obnetwork

27. Littlejohns TJ, Henley WE, Lang IA, et al. Abstract: *Vitamin D and the risk of dementia and Alzheimer disease.* Published online before print 8-6-14. Neurology. Org. http://www. doi.org/10.1212/WNL.0000000000000755 Neurology 10.1212/WNL.0000000000000755

28. Belluck P. *Protein May Hold the Key to Who Gets Alzheimer's.* The New York Times, 3-19-14. http://www.nytimes.com/2014/03/20/health/fetal-gene-may-protect-brain-from-alzheimers-study-finds.html?_r=0

29. Lu T, Aron L, Zullo J, et al. Abstract: *REST and stress resistance in ageing and Alzheimer's disease.* Published in Nature 507(7493): 448-54, 3-27-14. http://www.ncbi.nim.nih.gov/pubed/24670762

30. Fischer K. *Brains Need REST to Protect Against Alzheimer's Disease.* Healthline News, 3-20-14. http://www.healthline.com/health-news/aging-regulator-protein-could_-protect-against-alzheimers-032013

31. Kolata G. *Alzheimer's Tied to Mutation Harming Immune Response.* The New York Times, 11-14-12. Nytimes.com/2012/11/15/.../gene-mutation-that-hobbles-immune-response-is-linked to-alzheimers.html

32. Miller, T. *Scientists find 11 more genes linked to Alzheimer's disease.* New York Daily News, 10-28-13. www.nydailynews.com/life-style/health/scientists-find-11-genes-linked-alzheimer-disease-article-1.1499015

33. Alzheimer's Disease Education and Referral Center. *Identifying the Genetics of Alzheimer's Disease.* http://www.nia.nih.gov/alzheimers/publication/2013-2014-alzheimers-disease-progress-report/identifying-genetics-alzheimers

34. Hamilton J. *Gene Linked to Alzheimer's Poses a Special Threat to Women*. Shots-Health News: NPR, 4-14-14. http://www.npr.org/blogs/health/2014/04/14/3019988330/gene-linked-to-alzheimers-poses-a-special-threat-to-women
35. Weber B. *Alzheimer's: resveratrol targets certain protein interactions*. Medical News Today, 10-22-13. http://www.medicalnewstoday.com/articles/267776.php
36. Woerner A. *New Alzheimer's discovery could hold key to preventative treatments for high-risk patients*. Fox News, 10-22-13. www.foxnews.com/health/2013/10/22/new-alzheimers-discovery-could-hold-key-to-preventive-treatments-for-high/?intcmp=obnetwork
37. Alzheimer's Reading Room. *Family History a Known Risk Factor for Alzheimer's*. http://www.alzheimersreadingroom.com/2013/04/family-history-known-risk-factor-for.html
38. Science Daily. *Two parent with alzheimer's disease? Disease may show up decades early on brain scans*. Science Daily, 2-12-13.. <www.sciencedaily.com/releases/2014/02/14021214308.htm>.
39. Vassilaki M, Aakre JA, Cha RH, et al. Abstract *Multimorbidity and Risk of Mild Cognitive Impairment*. Article published in J Am Geriatr Soc 63(9):1783-90, 2015. http://www.ncbi.nih.gov/pubmed/26311270
40. Research: Most current research and studies on Prevagen https://www.prevagen.com/research/
41. Eisen, M. *The FDA looks into Quincy Bioscience claims for Prevagen*, Isthmus/Madison, Wisconsin, 12-5-13. http://www.isthmus.com/news/the-fdaa-looks-into-quincy-biosciences-claims-for-prevagen/
42. Truth in Advertising. *Prevagen*. 2-19-15. https://www.truthinadvertising.org/prevagen/
43. National Council Against Health Fraud. *Class-action suit filed against "memory supplement" marketers*. Consumer Health Digest #15-06; Your Weekly Update of News and Reviews. 2-8-15. http://www.ncahf.org/digest15/15-06.html
44. Mundasad, S. BBC News. *Brain may 'compensate' for Alzheimer's damage*. 9-14-14. http://www.bbc.com/news/health-29181843
45. Dallas ME. *Not Everyone with Alzheimer's Linked Protein Develops Dementia: Study* Health Day, 9-15-14. http://consumer.healthday.com/senior-citizen-information-31/senior-citizen-news-778/not-all-older-adults-with-alzheimer-s-related-protein-will-develop-...
46. Eisenberg, R. *Retiring Later Could Help You Fend Off Alzheimer's*. Forbes, 7-15-13. www.forbes.com/sites/nextavenue/2013/07/15/retiring-later-could-help-you-fend-off-alzheimers/
47. Emling S. *Speaking Two Languages May Slow Brain Aging*. The Huffington Post, 6-2-14. http://www.huffingtonpost.com/fifty

48. Castillo M. *Reading, writing may help preserve memory in older age.* CBS News. July 4, 2013 www.cbsnews.com/8301-204_162-57592342/reading-writing-may-help-preserve-memory-in older-age/
49. Payne C. *Brain Stimulation at any age may slow memory decline.* USA Today, 7-4-13. STORY: Healthy lifestyle habits may improve your memory, too http://www.usatoday.com/story/news/nation/2013/06/01/healthy-lifestyle-memory-adults/2371249/
50. Institute of Medicine. *Cognitive Aging: Progress in Understanding and Opportunities for Action.* Report Brief. April 2015 www.iom.edu/cognitiveaging
51. Alzforum. *Synaptic Function in Aging and AD.* 11-5-06. http://www.alzforum.org/news/conference-coverage/synaptic-function-aging-and-ad
52. Sifferlin, A. *Eating Fish Makes Your Brain Healthier, Study Says.* Time, 8-4-14. http://time.com/3079816/fish-brain-gray-matter-study-omega-fatty-acids/
53. *Fish Oil Boost Brain Power,* 7-16-14. https://www.yahoo.com/health/fish-oil-boosts-brain-power-says-science-91965585587.html
54. Reddy S. *A Diet Might Cut the Risk of Developing Alzheimer's.* The Wall Street Journal, 4-20-15. http://www.wsj.com/articles/a-diet-might-cut-the-risk-of-developing-alzheimers-1429569168?ru=yahoo?mod=yahoo_itp
55. Norton A. *Could Red Wine Ingredient Affect Progression of Alzheimer's.* Health Day, 9-11-15 http://consumer.healthday.com/cognitive-health-information-26/alzheimer-s-news-20/could-red-wine-ingredient-affect-progression-of-alzheimer-s-703195.html
56. Weintraub K. *Lifestyle changes are key to easing Alzheimer's risk.* USA Today, 7-14-14. http://usat.ly/1mBVBzH
57. Today's Geriatric Medicine. *For People With Dementia, Cataract Surgery Improves Vision, Cognition, Quality of Life.* http://www.todaysgeriatricmedicine.com/news/072514_news.shtml
58. Gill SS, Seitz DP. *What Older Individuals Can Do to Optimize Cognitive Outcomes.* Editorial, JAMA 314(8):774-775, 8-25-15
59. Chuang Y-F, An Y, Bilgel M, et al. Abstract : *Midlife adiposity predicts earlier onset of Alzheimer's dementia, neuropathology and presymptomatic cerebral amyloid accumulation.* Full article in Molecular Psychiatry,9-1-15. Doi:10.1038/mp.2015,129 http://www.nature.com/mp/journal/vaop/ncurrent/full/mp2015129a.html
60. Kunkle F. *Too much TV could raise the risk of Alzheimer's, study suggests.* The Washington Post, 7-20-15
61. Vemuri P, Lesnick TG, Przybelski SA, et al. Abstract: *Association of Lifetime Intellectual Enrichment With Cognitive Decline in the Older Population.*.Article published in JAMA Neurol. Published online on June 23, 2014. doi:10.1001/jamaneurol.2014.963

http://archneur.jamanetwork.com/article.aspx?articleid=1883334&result
Click=3&utm_medium=BulletinHealthCare&utm_term=062414&...

Chapter 5

1. Holland, K. *The Facts About Alzheimer's: Life Expectancy and Long-Term Outlook.* Healthline, 12-6-13. http://www.healthline.com/health/alzheimers-disease/life-expectancy#AverageLifeExpectancy3
2. Alzheimer's Disease Education and Referral Center. *About Alzheimer's Disease: Alzheimer's Basics.* https://www.nia.nih.gov/alzheimers/topics/alzheimers-basics
3. Health in Aging.Org. *Nursing Homes: Basic Facts and Information.* http://www.healthinaging.org/aging-and-health-a-to-z/topic:nursinghomes/
4. Cadigan RO, Grabowski DC, Givens JL, et al. Abstract : *The Quality of Advanced Dementia Care in the Nursing Home : The Role of Special Care Units.* Article published in Med Care 50(10): 856-62, 2012. http://www.ncbi.nim.nih.gov/pmc/articles/PMC3444818/
5. Kane RL, Finding the Right Level of Posthospital Care: *"We Didn't Realize There Was Any Other Option for Him."* Clinician's Corner, JAMA 35(3):284-93, 1-19-11
6. Gilley DW, Bienias JL, Wilson RS, Bennett DA, et al. Abstract : *Influence of Behavioral Symptoms on rates of institutionalization for persons with Alzheimer's Disease.* Cambridge Journals. Article published inPsychological Medicine/Volume/Issue 06: 1129-35, 2004. http://journals.cambridge.org/action/display/Abstract?fromPage=online&aid=242061&fileid=S0033291703001831
7. White MC. *Americans Are Totally Unprepared for This Shock.* Time, 6-26-14. http://time.com/2923939/retirement-2/
8. New York Nursing Home Abuse Lawyer Blog. *Video Camera Captures Two Nurse Aides Taunting and Abusing Elderly Dementia Patient.* 6-2-14 http://www.newyorknursinghomeabuselawyerblog.com/2014/06/video-camera-captures-two-nurs.html
9. Fox News.com. *Nursing home striptease was elderly residents' idea, lawyer says.* 4-9-14 http://www.foxnews.com/us/2014/04/09/lawyer-says-new-york-nursing-home-residents-voted-to-bring-in-strippers/?intcmp=latestnews
10. Hoffman J. *Watchful Eye in Nursing Homes.* The New York Times, 11-18-13 http://well.blogs.nytimes.com/2013/11/18/watchful-eye-in-nursing-homes/?php=true&_type=blogs&_r=0
11. Lester K. *Madigan proposes allowing cameras at nursing homes.* Washington Times, 9-8-14. http://www.washingtontimes.com/news/2014/sep/8/new-proposal-would-allow-cameras-at-nursing-homes/
12. Allen M. *One Third of Skilled Nursing Patients Harmed in Treatment.* ProPublica, 3-3-14. http://www.propublica.org/article/one-third-of-skilled-nursing-patients-harmed-in-treatment

13. Centers for Medicare & Medicaid Services. Medicare & You 2016 Page 41
14. Kotz D. *Study questions drugs given to many with advanced Alzheimer's*. Boston Globe, 9-8-14
15. Healy M. *Dementia patients continue to get medications with little, no benefit*. LA Times, 9-9-14. http://www.latimes.com/sciencenow/la-sci-sn-dementia-medications-benefit-20140908-story.html
16. Katz K. *Nursing Credentials 101: From LPN & LVN to BSN & DNP*. 12-10-13. http://www.rasmussen.edu/degress/nursing/blog/nursing-credentials-101-from-lpn-lvn-bsn-dnp/
17. Howard MH. *Difference Between a CNA, LPN & RN*. eHOW http://www.ehow.com/about_6554048_difference-between-cna_lpn-rn.html
18. Pfeiffer S, Jolicoeur L. *Study: Late-Stage Dementia Patients "Slammed Around Health Care System."* 10-19-11. www.wbur.org/2011/10/19/alzheimers-end-life-care
19. Ouslander JG, Berenson RA. *Reducing Unnecessary Hospitalizations of Nursing Home Residents*. N Engl J Med 365: 1165-67; 9-29-11. http://www.nejm.org/doi/full/10.1056/NEJMp1105449
20. Teno JM, Mitchell SL, Kuo SK, et al. *Decision-Making and Outcomes of Feeding Tube Insertion: A Five-State Study*. Journal of the American Geriatrics Society, 59: 881-86, doi: 10.1111/j.1532-5415.2011.03385.x
21. John Hancock. *John Hancock National Study Finds Long-Term Care Costs Continue to Climb Across All Provider Options*. 7-30-13. http://www.johnhancock.com/about/news_details.php?fn=jul3013-text&yr=2013
22. National Association of Insurance Commissioners. Consumer Alert. *Long-Term Care insurance: What You Should Know*. http://www.naic.org/documents/consumer_alert_ltc.htm
23. Alzheimer's Association. *2014 Alzheimer's disease facts and figures*. http://www.sciencedirect.com/science/article/pii/S1552526014000624
24. Galewitz P. *States Accelerate Shift of Nursing Home Residents into Medicaid Managed Care*. Kaier Health News, 2-11-14. http://www.kaiserhealthnews.org/Stories/2014/February/11/states-move-faster-on-moving-nursing-home-residents-to-medicaid-managed-care.aspx
25. Bernstein N. *Pitfalls Seen in a Turn to Privately Run Long-Term Care*. The New York Times, 3-6-14. http://www.nytimes.com/2014/03/07/nyregion/pitfalls-seen-in-tennessees-turn-to-privately-run-long-term-care.html?ref=nyregion&_r=0
26. Kopke S, Muhlhauser I, Gerlach A, et al. *Effect of a Guideline-Based Multicomponent Intervention on Use of Physical Restraints in Nursing Homes*. JAMA 307(20): 2177-184, 2012
27. Small GW. Editorial: *Treating Dementia and Agitation*. JAMA 311(7):677-78, 2-19-14

28. Nursing Home Abuse Guide.org. *Chemical restraints on Elderly* http://www.nursinghomeabuseguide.org
29. Kuehn BM. *APA Targets Unnecessary Antipsychotic Use.* JAMA 310(18): P.1909-10, 11-13-13
30. FDA Alert. Information on Conventional Antipsychotics. http://www.fda.gov/drugs/drugsafety/postmarketdrugsafetyinformationforpatientsandproviders/ucm107211.htm
31. Francis E, Feeley J, Voreacos D. *J & J Ordered to Pay $1.1 Billion Penalty Over Risperdal.* Bloomberg, 4-12-12. www.bloomberg.com/news/2012-04-11/jnj-told-to-pay-1-1-billion-penalty-in-arkansas-risperdal-trial.html
32. Voreacos D, Fisk MO . *J & J Will Pay $181 Million to Settle Risperdal Ad Claims.* Bloomberg, 8-30-12. www.bloomberg.com/news/2012-08-30/j-j-will-pay-181-million-to-settle-risperdal-ad-claims.html
33. Thomas K. *Johnson & Johnson Unit Settles State Cases Over Risperdal.* TheNew York Times, 8-30-12. www.nytimes.com/2012/08/31/business/johnson-johnson-unit-settles-state-cases-over-risperdal.html
34. Thomas K. *J & j to Pay $2.2 Billion in Risperdal Settlement.* New York Times, 11-4-13. www.nytimes.com/2013/11/05/business/johnson-johnson-to-settle-risperdal-improper-marketing-case.html?pagewanted=1&_r=0
35. Thomas K. *Arkansas Court Reverses $1.2 Billion Judgment Against Johnson & Johnson.* The New York Times,3-20-14. http://www.nytimes.com/2014/03/21/business/arkansas-court-reverses-1-2-billiion-judgment-against-johnson-johnson.htmll?_r=1
36. Haelle T. *Certain Antipsychotic Meds Tied to Kidney Problems in Elderly.* HealthDay, 8-19-14. http://consumer.healthday.com/senior-citizen-information-31/misc-aging-news-10/certain-antipsychotic-meds-tied-to-kidney-woes-in-elderly-study-says-690863....
37. U.S. Dept of Health and Human Services/National Institutes of Health. *Antipsychotic Medications Used to Treat Alzheimer's Patients Found Lacking.* 10-11-06. www.nih.gov/news/pr/oct2006/nimh-11.htm
38. Harris G. *Antipsychotic Drugs Called Hazardous for the Elderly.* The New York Times, 5-9-11. http://www.nytimes.com/2011/05/10/health/policy/10drug.html?_r=1&scp=1&sq=use%20...
39. Kuehn BM. *Efforts Stall to Curb Nursing Home Antipsychotic Use.* Medical News & Perspectives JAMA 310(11): 1109-10, 9-18-13
40. Mitka M. *CMS Seeks to Reduce Antipsychotic Use in Nursing Home Residents With Dementia.* Medical News & Perspectives. JAMA 308(2): 119-21. 7-11-12
41. Porteinsson AP, Drye LT, Pollock BG, et al. *Effect of Citalopram on Agitation in Alzheimer Disease : The CitAD Randomized Clinical Trial.* Original Investigation. JAMA 2014: 311(7): 682-691. doi:10.1001/jama.2014.93

42. Cummings JL, Lyketsos CG, Peskind ER, et al. *Effect of Dextromethorphan-Quinidine on Agitation in Patients With Alzheimer Disease Dementia: A Randomized Clinical Trial.* Original Investigation. JAMA 314(12): 1242-54, 2015
43. Thomas K. *In Race for Medicare Dollars, Nursing Home Care May Lag.* The New York Times/Business Day http://www.nytimes.com/2015/04/15/business/as-nursing-homes-chase-lucrative-patients-quality-of-care-is-said-to-lag.html?_r=0
44. Thomas K. *Medicare Star Ratings Allow Nursing Homes to Game the System.* The New York Times, 8-24-14. http://www.nytimes.com/2014/08/25/business/medicare-star-ratings-allow-nursing-homes-to-game-the-system.html?ref=business
45. Thomas, K. *Government Will Change How it Rates Nursing Homes.* The New York Times, 2-12-15. http://www.nytimes.com/2015/02/13/business/government-will-change-how-it-rates-nursing-homes.html?_r=0
46. Thomas K. *Medicare Toughens Standards on Nursing Homes.* The New York Times/Business Day, 2-20-15. http://www.nytimes.com/2015/02/21/business/nursing-home-ratings-fall-as-tougher-standards-take-effect.html?ref=health&_r=0

Chapter 6

1. Fox News. DC duel: *Seth Rogen blasts senators for skipping his testimony; Ben Affleck plays to packed house.* 2-27-14. http://www.foxnews.com/entertainment/2014/02/27/seth-rogen-talks-about-alzheimers-disease-at-senate-hearing-calls-out-senators/?intcmp=features
2. Kenney L. *Seth Rogen Details His 'Family Love Story' And Mother-in-Law's 'Brutal' Alzheimer's Battle.* https://www.yahoo.com/health/seth-rogen-details-his-family-love-story-and-105007774542.htmlwww.usatoday.com/story/news/nation/2013/03/19/dementa-alzheimers-medicare/1994951/
3. Reid TR. *Where's the War on Alzheimer's?* AARP Bulletin. Jan-Feb 2015, Vol 56 No1
4. Lloyd J. *One in three elderly have dementia when they die.* USA Today. 3-19-13.
5. Avorn J, Kesselheim AS. *The 21st Century Cures Act – Will It Take Us Back in Time?* N Engl J Med 372: 2473-75, 6-25-15. http://www.nejm.org/doi/full/10.1056/NEJMp1506964
6. Zuckerman DM, Jury NJ, Silcox CE. *What would impact of 21st Century Cures Act be on your healthcare costs and the lives of Alzheimer's patients?* National Center for health research, 11-23-15. http://center4research.org/public-policy/21st-century-cures-act-and-similar-policy-efforts-at-what-cost/
7. Zuckerman DM, Jury NJ, Silcox CE. *21st Century Cures Act and similar policy efforts: at what cost?* http://www.bmj.com/content/351bmj.h6122
8. State Alzheimer's Disease Plan Resource Center. *The National Alzheimer's Project Act (NAPA).* http://napa.alz.org/national-alzheimer's-project-act-backgroun

9. Alzheimer's Association/Advocacy. *Public Policy Victories* http://www.alz.org/advocacy/victories.asp
10. Antlfinger, Carrie. *Aiming for a 'dementia friendly' society*. The Tampa Tribune, 9-27-15, P. 17 TBO.com
11. Agnvall, Elizabeth. *Making a Town Dementia-Friendly*. AARP Bulletin, Jan-February 2014, P.7 Aarp.org/bulletin
12. Ertel KA, Glymour M and Berkman LF. *Effects of Social Integration on Preserving Memory Function in a Nationally Representative US Elderly Population*. American Journal of Public Health 98(7): 1215, July 2008. Published ahead of pprint on May 29, 2008 as 10,2105/AJPH.2007.113654. The latest version is at Http://www.iph.org/cgi/doi/10.2105/AJPH.2007.113654
13. Ostrow N. *Alzheimer's Leaves Patients, Caregivers Feeling Isolated*. Bloomberg News. www.bloomgerg.com/news/2012-09-20/alzheimer-s-leaves-patients-caregivers-feeling-isolated.html
14. National Institute on Aging/ Alzheimer's Disease Education and Referral Center. *Alzheimer's Disease: Unraveling the Mystery*. Publication date: Sept 2008. Page last updated: 1-22-15. http://www.nia.nih.gov/alzheimers/publication/part-2-what-happens-brain-ad/changing-brain-ad
15. Associated Press. *Alzheimer's patients get buddies*. The Tampa Tribune, 3-16-14. P.21 TBO.com
16. Associated Press. *More Alzheimer's patients finding care far overseas*. Fox News. 12-30-13. www.foxnews.com/health/2013/12/30/more-alzheimers-patients-finding-care-far-overseas/

www.ingramcontent.com/pod-product-compliance
Lightning Source LLC
Chambersburg PA
CBHW021943170526
45157CB00003B/906